Lecture Notes in Earth Sciences   96

Editors:
S. Bhattacharji, Brooklyn
G. M. Friedman, Brooklyn and Troy
H. J. Neugebauer, Bonn
A. Seilacher, Tuebingen and Yale

W0106554

Springer-Verlag Berlin Heidelberg GmbH

John Cobbing

The Geology and Mapping of Granite Batholiths

With 40 Figures

 Springer

Author

Dr. John Cobbing
25, Main Road
NG12 2BE Radcliffe on Trent
Great Britain

"For all Lecture Notes in Earth Sciences published till now please see final pages
of the book"

ISSN 0930-0317

ISBN 978-3-540-67684-3   ISBN 978-3-540-45055-9 (eBook)

DOI 10.1007/978-3-540-45055-9

Cataloging-in-Publication data applied for
Die Deutsche Bibliothek - CIP-Einheitsaufnahme
Cobbing, John: The geology and mapping of granite batholiths / John Cobbing.

        (Lecture notes in earth sciences; 96)
   ISBN 978-3-540-67684-3

© Springer-Verlag Berlin Heidelberg 2000

Originally published by Springer-Verlag Berlin Heidelberg New York in 2000

Typesetting: Camera ready by author
Printed on acid-free paper   SPIN: 10711221   32/3130-543210

Preface

This book is mainly about the field geology of granites at all scales from that of a single outcrop to plutons and batholiths. All field geologists work initially at the scale of the outcrop, consequently most of the phenomena treated herein are those which are visible at outcrop scale. However, granites typically occur as plutons and batholiths, some of which are so large as to apparently defy any effort at systematic treatment. Having had the opportunity of mapping two very large and very different batholiths, namely the Coastal Batholith of Peru and the tin granites of Southeast Asia, I have found that it is possible to map large batholiths within a relatively short time, so that the geology of the batholith as a whole can be appreciated. Moreover batholiths are one of the most common modes of granite occurrence, so it makes sense to study them at their natural scale.

During my working life I have worked with many geologists from underdeveloped countries and this book is mainly to help them in unravelling the geology of their native batholiths.

I have been lucky with my friends and colleagues of many nationalities, and I particularly thank Wallace Pitcher, who took me on as an untried apprentice in Peru, and who, by his kindness and example, showed me how to look at granites properly.

Some geologists may consider that too much of this book is devoted to stating the obvious. It is written mainly for those many third world geologists engaged in mapping granites in inaccessible terrains, with inadequate logistic support and an inadequate time frame in which to complete their task.

The structure of the book has been devised with these requirements in mind and is based on workshops given in Malaysia, Indonesia, China and Norway. It is concerned principally with the practicalities of field mapping but the first four chapters are of a more academic turn, and are meant to provide an outline of the current state of theoretical knowledge of granite geology as the background to field work. The bulk of the remainder is concerned with the purely practical aspects of the regional mapping of granitic terrains, and the final chapters provide guidelines for the requirements to be borne in mind when preparing geological maps of these terrains, and the presentation of the data in written form.

I have tried to make the descriptive sections on granite geology as comprehensive and intelligible as possible. Most of the phenomena described are not now controversial, but in those which still are, I have endeavoured to indicate where the areas of controversy lie, and to state my own opinion if I have one. I have not attempted to provide a fully comprehensive account of every aspect of granite geology, but have tried to cover most of the features which are likely to be met with by field geologists. Paradoxically regional studies of granites are best served by the very careful observation of the details of granite geology at the scale of the outcrop. I have endeavoured to focus on these details and show how they may be systematically identified and recorded. This will ensure that the regional interpretation is solidly based on an accurate and comprehensive record of the field geology. All granites result from the interaction of different processes during their generation, crystallisation and emplacement, and many of these processes leave a permanent imprint on the rock which can, with care, be identified in the field. One of the objectives of this book is to indicate how the details of the field geology provide information relating to these processes.

Many granites are deceptively simple in their appearance and there are many examples of granites which have been studied carefully and in great detail, but which have continued to yield additional, and sometimes quite contradictory information to later investigators. The history of granite geology is full of instances of such reversals of interpretation. With this background it is unlikely that any study will not be subject to reinterpretation at some stage. This should not however discourage future work, for there are still many granite terrains of which we remain wholly ignorant, and which need to be mapped in order to provide the basis for comparison and further work. The people who do the regional work form that section of the profession to which this book is directed, and although they are particularly vulnerable in this respect, since the reconnaissance nature of their task renders their conclusions subject to later revision, they should take heart and persist. Their work provides the basis for all subsequent investigations.

Finally I thank all my friends and colleagues and especially Peter Pitfield, Don Mallick, Fiona Darbyshire, Michael Schwartz, Bernd Lehmann, Bob Beckinsale, Willy Taylor, Michael Atherton, Andrew Bussell, Paul Bateman, Gordon Gastil, Cesar Vidal, Julio Caldas, Julio Garayar, Wilfredo Sanchez, Bill Mc Court, Michael Crow, John Aspden, Martin Clarke, Teoh Lay Hock, Charles Hutchison, Hong Dawei, Somboon Khositanont, Somchai Nakapadungrat, Oystein Nordgulen, Brian Sturt, Larissa Dobrezhinetskaya, Valery Vitrin, Carmen Galindo and Cesar Casquet. I would also like to pay a tribute to the late Chamrat Mahawat who was my good friend, and who made it possible for me to work in Thailand.

ACKNOWLEGEMENTS

I thank Dr Reedman of the International Division of BGS for his encouragement and help towards the production of this book.

Table of Figures

Contents

1 Introduction

The average composition of the continental crust is granodioritic (Clarke, 1991) and in some orogenic belts granites form as much as 30% of the surface outcrop and are an essential component of the continental crust. Although they occur in great belts which commonly extend for hundreds or even thousands of miles, each belt is made up of individual bodies,which may be very large, or quite small, and which are invariably rather complex. Until recently granite studies were mainly pursued on such individual bodies by individual geologists resulting in a patchwork of information which, while excellent in itself, has merely dealt with the tip of the iceberg in relation to the problem as a whole. Understanding the geology of our globe requires a much more comprehensive picture of the granites, which form so large proportion of the crust. This can only be achieved by studying them in a regional way, and is all the more necessary since granites result from plate tectonic processes which both produce new crust, and recycle older crust.

Several regional studies have been conducted, notably in Peru (Pitcher et al., 1985), the Tin Belt of Southeast Asia (Cobbing et al., 1992), the Sierra Nevada of the western USA (Bateman, 1992), Baja California (Gastil et al., 1975) the Lachlan Belt of Australia (Chappell & White, 1992) and Southeast China (Xu Keqin et al., 1980). Nevertheless there are still enormous belts of granite which remain totally or partially unknown, and if this is to be remedied regional mapping programmes will have to be implemented.

Granites are difficult rocks to work with. Unlike sedimentary or volcanic formations, they do not occur as well defined bands with a known top and bottom, which can be followed for long distances and traced round complex structures. They occur as separated bodies, each of which occupies a particular section of the earth's crust. Most of them have no discernible base or top and are internally complex, so that in order to map them it is necessary to be familiar with the whole range of their geological attributes, and to adopt field procedures which will ensure that these are all correctly recorded.

In any study of granitic rocks it is now customary to employ methods of geochemical and isotopic analysis, and accordingly the field mapping should be oriented towards that objective from the outset. It is only comparatively recently that analytical techniques have advanced to the stage where fully integrated regional studies have become a real possibility. The purpose of this work is to advocate the undertaking of such studies, and also to provide an account of the appropriate field methods by which they can be done.

J. Cobbing: LNES 96, pp. 1–4, 2000.

All geological investigations depend on the quality of the field observations, otherwise the resulting analytical data are seriously flawed. The methods for regional mapping which are presented here have been developed over many years in inaccessible and inhospitable terrains. While they are effective in these conditions they can also be readily adapted to more congenial areas.

For historical reasons the European tradition of granite investigation has been to map every pluton in the greatest possible detail. In Europe the proportion of granite to the total surface area is generally rather low and much of it is not well exposed, so that it makes sense to look carefully at every exposure. In some other regions the adoption of modern scientific methods is quite recent and in some cases has been superimposed upon a quite different cultural tradition. Moreover in some areas the proportion of granite to surface area is extremely high and they are so well exposed that there is no possibility of looking at every outcrop. In such places a tradition of regional mapping has developed. One result of these contrasting approaches was that geologists working in the Americas, China and Australia developed clearer concepts of the regional development of particular kinds of granitic suites.

Before the development of plate tectonic theory geologists realised that the existing geosynclinal model did not fully account for the observed differences in granite geology. For example there did not seem to be a satisfactory explanation for the existence of tin granites in some orogenic belts as opposed to base metal mineralised granites in others. Perhaps an exception was the early recognition that the alkali granites of continental interiors and of the Oslo graben were fault controlled in their anorogenic or rift related environments. Partially as a result of this inadequacy the plate tectonic concept, as presented in the form of the Wilson Cycle was generally welcomed as providing a potential solution to many of these underlying problems.

Various plate tectonic situations are now recognised, wherein certain combinations of volcanic and plutonic rocks of particular compositions are considered to be representative of specific tectonic settings developed at different stages of a Wilson cycle. While there is still much unresolved complexity the theory still provides the framework within which most granite geologists work.

The great virtue of the theory is that it has shown the necessity for studying geological phenomena in both a regional and detailed way, and it is fortunate that new technological advances, such as satellite imagery and automated analytical systems, have rendered such studies possible. Fieldwork, however cannot be greatly speeded up, and in order to cover larger areas the observations have to be more thinly spread and of the highest possible quality, which, under some conditions is not always easy.

All geological phenomena are characterised by a virtually infinite degree of complexity requiring investigation at all scales. However, in regional work it is impossible to study all the granites of an orogenic belt in detail and choices are made on the basis of the best information available. If, in a batholith there are a thousand plutons but only ten have been mapped, it will be one of those ten which will benefit from the first geochemical study. Modern mapping programmes should endeavour to avoid this kind of imbalance. The essential objective of regional study is to spread high quality information as evenly as possible so as to be representative of the larger region. Whether a geologist is mapping at a scale of 1:10 000 or 1:500 000 the situation is always the same, the geologist is looking at and describing an outcrop. There is no reason for the quality of observation to be better at one scale rather than the other. The objective is to ensure that all plutons studied in a regional way can be treated on an equal basis. It is possible for a few people to map complete belts of granite in a relatively short time and provide a satisfactory and useful, if not perfect understanding of the granites in the belt. A regional framework can be established within which every granite and its variants can be placed, and which will provide the basis for more detailed studies of particular problems. Properly conducted regional studies will normally identify those plutons which are of special interest, in that their future study may shed light on some intractable area of granite geology.

Modern technology ensures that virtually everything one needs to know about granites can now be determined by petrological, geochemical or isotopic methods, sometimes on single crystals, and it is possible to envisage a time when some of the outstanding problems will have been resolved by this approach. But each granite is an expression of a hierarchical scale of attributes from molecular–crystal–grain aggregate–pluton–batholith–granite belt. Each of these scales has its own frame of phenomenological reality and all deserve to be studied equally. The newest techniques, by their very nature, are focussed towards the scale of the single pluton, or even the scale of the single crystal. This fine focus approach very often yields results of regional significance which, paradoxically may sometimes be best understood within a regional framework should that be available. Thus the first stage of research must be the production of a map showing the regional extent and geological relationships of all the granites.

Granite belts are widespread in the continental crust and are characterised chiefly by their diversity. They can be of any age from Archaean to Tertiary and have either an extended or restricted compositional range. Their outcrop form and compositional character is equally diverse. Some are long and thin like those of the Andes, while others like the Lachlan belt of Australia and the granites of Southeast China, are much broader in proportion to their length. In contrast the belts of Proterozoic Rapakivi granites in Finland and Russia form a distinct class not precisely matched by later granites and certainly all of the Phanerozoic belts of Europe are distinct from one another.

In short the diversity of granites which one might expect to see within a single granite belt is more than matched by that between the belts themselves. The reasons for this degree of diversity remain obscure and will remain so until enough regional work has been done to establish the nature of these differences. We can be sure that they will not be simple, but may in all probability be a subtle combination of source region, varying degrees of mantle and crustal evolution, perhaps in combination with structural factors associated with different stages of the Wilson Cycle.

This diversity is the more remarkable because all these rocks are granitoids which plot together on a Streckeisen diagram and can be confidently classified as granite because of their evident similarity. It is this strange combination of unity and diversity which, among other things, has led to the various controversies which have arisen among geologists as to the origin and status of these rocks.

2 Systems of Classification

Systems of classification for granitic rocks can be divided into two main classes, those which are compositionally based and those which are not directly related to composition. It can be stated as a general principle that compositionally based systems are to be preferred, since they are more objective in their methods of determination, being entirely dependent on the measurement of mineralogical or geochemical data. However, alternative systems provide useful supplementary information on the wider relationships of granitic rocks and the geological situations in which they occur.

The main systems of classification based on mineralogy, lithology and major element geochemistry, which have been long established, are still in common use and provide the empirical base for all subsequent classifications. However, they are commonly used in combination with other, structural or tectonic qualifying criteria, which have been considered to reflect some aspect of the nature of granite genesis or of pluton emplacement. Structurally or tectonically derived nomenclatures, such as syn- or post-tectonic, post-orogenic and anorogenic, have been used for many years and are still commonly employed. Plate tectonics has added to the list and we now have oceanic arc, volcanic arc, collisional, post-collisional and within plate granites. Some of these are clearly synonymous with earlier terminologies but others are not. The end result is that granites and granite belts are sometimes described by inappropriate or misleading terms. Consequently, the following discussion will attempt to indicate those situations where they can be usefully employed, or conversely, dispensed with.

Since unrelated criteria have commonly been used together, a brief comment on the nature of geological classifications may not be out of place. The necessity for classification has been a distinguishing feature of the natural sciences since their inception and their use enables scientists to establish order within the apparent confusion of nature. Historically geology has been, and continues in large measure to be, a science of observation, followed by interpretation. Analysis of the data leads to the development of hypotheses or models, which can be tested by further observations or experimental procedures. This results in understanding of the process which gave rise to the observed phenomena. There is consequently a clear line of development from empirical description to a process oriented or genetic terminology. The history of granite classification is an illustration of the permanent state of tension between these two poles. A good classification should be simple, objectively based and reflect geological reality. Above all it should be useful, and capable of illuminating problem areas. It should not provide a mental straitjacket which precludes thought.

J. Cobbing: LNES 96, pp. 5–16, 2000.
© Springer-Verlag Berlin Heidelberg 2000

Because of the complexity of the processes resulting in the generation and emplacement of granites it is vital that the initial classification should be as empirical as possible. This means that it should be immediately useful in the field and so focus on the identification and description of the constituent minerals, the interaction of their grain boundaries and the resulting granularity and texture, followed by observations on structural and other phenomena. The field description is followed by microscopic study, geochemical analysis of selected samples for major and minor elements, and isotopic analysis of more rigorously chosen material. By following this procedure a body of data is progressively assembled which is as empirically based as it is possible to be, allowing for human error, and which can provide a secure platform for subsequent interpretation. Other empirically based information such as the field recording of geophysical or radiometric data, or the separation and study of heavy minerals, broadens the base of the purely factual range of information which can be obtained from any granite.

In natural outcrops however, this doesn't often happen. Only a tiny proportion of all the granite mapped is ever subjected to the full gamut of analytical techniques now available. All the more reason then for those granites which do not qualify for the full treatment to be as carefully described as those which do. They are, after all, volumetrically more representative of the body than those which are analysed.

2.1
Empirical Classifications

2.1.1
Lithological

The lithological terms of diorite, tonalite, granodiorite and granite have been used for over a century (Rosenbusch, 1887, Johanssen, 1937, Hatch et al., 1951). These categories are determined by point counting of the modal proportions of quartz, K-feldspar and plagioclase in thin section. The method is completely empirical and simple, though the point counting of thin sections is time consuming and in the case of coarse rocks can give misleading results. This difficulty is resolved by counting points on large polished surfaces, which are sometimes stained to distinguish between K-feldspar and plagioclase. It is also possible to estimate the proportions of these components in the field and with practice this can become surprisingly accurate.

Because point counting is so laborious it is now common practice to determine the lithologies by plotting normative data directly on to a QAP triangular Streckeisen diagram. This can be misleading because perthitic alkali feldspar contains both orthoclase and albite. By the point counting method all the perthitic albite is included as K-feldspar whereas by plotting normative data the K_2O and Na_2O are separated, giving a result which is compositionally more accurate but

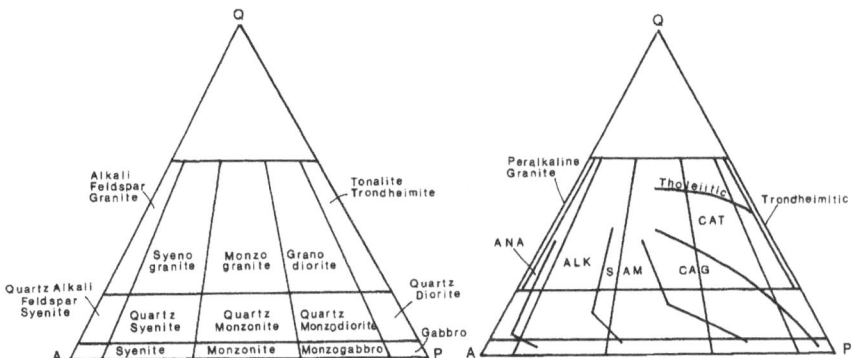

Fig. 1. QAP diagram (Streckeisen, 1976). A. Modal classification of granitoids. B. Granite lineages distinguished by Lameyre & Bowden (1982) CAT calc-alkaline tonalite, CAG calc-alkaline granodiorite, SAM subalkaline monzogranite, ALK aluminous potassic and ANA alkaline sodic

under-represents alkali feldspar. These difficulties are surmounted by using the Q' ANOR plot of Streckeisen & Le Maitre (1979) where Q'=(Q+Or+Ab+An) and the parameter ANOR=100 An/(An+Or).

An alternative method for overcoming these difficulties was developed by La Roche (1964, 1978) and modified by Debon and Le Fort (1983) in which the parameters are Q=Si/3–(K+Na+2Ca/3)and P=K–(Na+Ca). Although the latter diagram has begun to be used more frequently in recent publications, the QAP triangular diagram of Streckeisen has retained its popularity, and has even been pressed into service to distinguish between S and I–type granite (Bowden et al., 1984).

The versatility of the Streckeisen triangle has been explored by Lameyre and Bowden (1982) who have shown that different rock lineages may be identified which correspond to some naturally occurring granite populations. (Lameyre & Bowden, 1982). Bowden et al. (1984) distinguished five granite lineages on the Streckeisen triangle, CAT calcalkaline tonalitic or trondhjemitic, CAG calcalkaline granodiorite, SAM subalkaline monzonitic, ALK aluminous potassic and ANA alkaline soda. They also distinguished an area of 'crustal granites' and, in addition, areas which distinguished S–type, I–type and A–type granites, which, however, seem to be less useful in practice than the lineages. The versatility of the QAP diagram was further explored by Lameyre and Bonin (1992) who considered it to be 'a remarkable tool for distinguishing granite series even in the field'. These lineages correspond to distinctive examples of I–type or A–type plutonism and the reason for their ready identification on this simple diagram is that they have an extended compositional range from quartz diorite to granite. Conversely S–types are not well distinguished because they have a restricted range, normally only from granodiorite to granite. Moreover they plot in the narrowest part

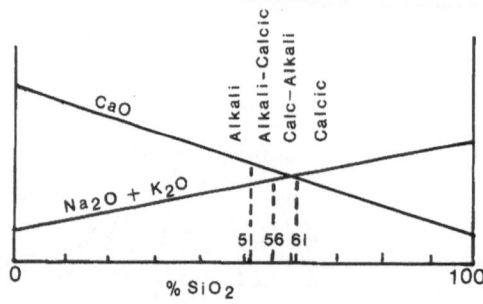

Fig. 2. The calc-alkali index of Peacock (1931)

of the diagram where there is insufficient space to portray such systematic variation.

2.1.2
Major Element Geochemistry

The calc-alkali index was defined by Peacock (1931) as those rocks in which Na_2O+K_2O exceeds CaO at SiO_2 values of between 55% and 61%. In those cases where these conditions obtained at higher silica values the rocks are classified as calcic, for values between 50% and 55% SiO_2 they are alkali calcic and for these below 50% they are alkalic. The calc-alkali index is useful for those granitoid suites with an extended compositional range such as those of the American cordilleras, but for those with a restricted range, where the lowest SiO_2 value is about 65%, the system cannot be applied with any confidence. There are many such belts, as for example, most of the tin granites, as well as those populations dominated by granodiorite–monzogranite, which many geologists consider to be the most characteristic granites of orogenic belts. In such cases other diagrams can be used to establish the calc alkali index. Those most commonly used are the K_2O+Na_2O vs SiO_2 diagram of Irvine & Barager (1971) and Kuno (1969) which define the fields of alkaline, calc-alkaline and tholeiitic granites.

A further method for determining the alkali balance was devised by Pecerillo and Taylor (1976) for use with volcanic rocks, but which has been widely used by authors writing on granite geology. It is particularly useful for distinguishing high K calc-alkali granites which, according to Roberts and Clemens (1993) are the characteristic granitic components of certain orogenic belts.

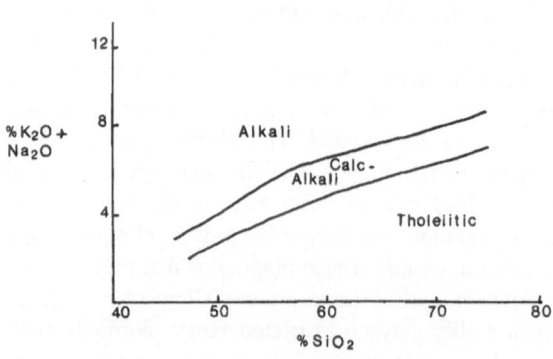

The concept of alumina saturation was defined by Shand (1927, 1947). Granites were said to be peraluminous, metaluminous or peralkaline according to the ratio of Alumina to Lime, Soda and Potash.

Fig. 3. Alkali discriminant diagram of Kuno (1969)

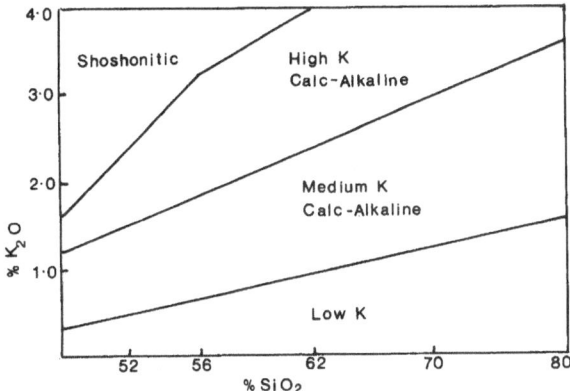

Fig. 4. K_2O vs SiO_2 discriminant diagram of Peccerillo & Taylor (1976)

Peraluminous	Metaluminous	Peralkaline
A/CNK>1	A/CNK<1	A/NK<1

The peraluminous index has attracted much attention since it was utilised by Chappell & White (1974) as one of the discriminants whereby S and I–type granites could be distinguished. While this claim has now been shown to be only partly true, the index continues to be of fundamental importance in granite geology. It has been incorporated by Debon & Le Fort (1983) into a comprehensive system of major element classification which combines the aluminous index in a bivariant plot against the parameter B=(F+Mg+T) which is proportional to the weight of dark minerals present. This is particularly useful because it reflects the major evolutionary trends of magmatic associations and clearly illustrates how metaluminous lineages converge with peraluminous ones by differentiation (Cobbing et al., 1992). However, the simplest way of comparing the aluminous values for different granite populations is by using binary plots of A/CNK vs SiO_2

These major element parameters of the alumina index and the calc-alkali index provide a powerful method of granite classification, especially when used in conjunction with the lithological system of Streckeisen, which not only classifies the rocks, but provides additional insights into their origin, lineage and pathways of differentiation (Lameyre & Bowden, 1982, Bowden et al., 1984).

2.1.3
Mineralogical

2.1.3.1
The Magnetite–Ilmenite Series

Ishihara (1977) divided granites into two classes based on their opaque mineral content which he designated magnetite series and ilmenite series granites. The opaque oxide mineralogy is controlled by the oxygen fugacity of the crystallising magma, magnetite series granites are oxidised and ilmenite series granites reduced. Magnetite series granites are associated with copper–base metal mineralisation and ilmenite series with tin–tungsten mineralisation. The system was developed principally as an adjunct to mineral exploration. Ishihara considered the ilmenite series granites to be reduced as a result of the incorporation of graphitic material from the source region into the crystallising magma and thus suggested that they were equivalent to the S–type granites of Chappell & White (1974) and that the magnetite series granites corresponded to their I–types. However matters are much more complicated than this and although S–type granites are virtually always of the ilmenite series, I–type granites may be either of the ilmenite or the magnetite series (Cobbing et al., 1992). Within some I–type belts distinct domains of magnetite and ilmenite series granites are present.

The great advantage of this system is that the magnetic properties of any granite can be directly measured in the field using a hand held Kappameter. Nearly always each granite body has a range of values which is distinctive and which in many cases helps to clarify problems of field identification. The routine use of such an instrument in the field is strongly recommended. It is also possible to use the instrument on old granite collections, thereby injecting new energy into old controversies.

2.1.3.2
Zircon

Pupin (1980) developed a system of classification which depended upon the separation of zircon crystals and the study of their morphology. Pupin was able to recognise seven distinct classes of granite by this method, which he interpreted as resulting from the influence of crustal, mantle and mixed crustal–mantle sources in their generation. The method cannot be faulted for empirical rigor but it is too specialised to be used in a routine way by most field geologists.

2.2
Typological Systems

The system was first enunciated by Chappell & White (1974) and subsequently modified by Pitcher (1979, 1983). Chappell & White (1974) distinguished S–type and I–type granites by their geological, geochemical and isotopic characteristics and related them to differences in their source regions. They considered that S–types were derived from a crustal sedimentary protolith and I–types from a crustal igneous protolith and that the composition of the source region was reflected in the composition of the granite. Subsequently White & Chappell (1983) accepted an additional A–type as first suggested by Loiselle and Wones (1979). In the case of the Lachlan Fold Belt, they proposed that this had resulted from magma genesis in a crustal igneous protolith which had been depleted as a result of a former I–type magma generation event. The refusion of the depleted protolith was thought to have resulted from the participation of fluorine-rich volatiles. Chappell & White consequently viewed their scheme as being strictly related to crustal source regions of different composition. They have never claimed that the typological scheme was related to tectonic setting. However, within the framework of plate tectonics, which is now our generally accepted guiding principle, and in the light of the successful identification of former tectonic settings by the geochemical characteristics of volcanic rocks erupted during their development (Pearce & Cann, 1973) it was only a matter of time before a connection between the typological scheme and tectonic setting was made.

Pitcher (1979, 1983) proposed that S–types or Hercynotypes as he called them, are typical of collisional settings and are tin associated. He divided I–types into two categories. Andinotype or Cordilleran I–types are those of the western Americas, with a compositional range from gabbro to granite and of predominantly tonalitic or granodioritic composition, the association being that of subduction at a continental margin. Other widely occurring I–types differ from Cordilleran I–types in having a bimodal expression being chiefly represented by high K monzogranites and granodiorites associated with smaller gabbroic bodies. Such granitoids occur in the Caledonides of Scotland and Ireland and were first designated by Pitcher (1979, 1983) as Caledonian I–types. Subsequently he linked them to a broader tectonic association which included post-collisional and post-orogenic situations, often in conjunction with uplift and the formation of molasse basins and andesitic plateau vulcanicity (Pitcher, 1993). I–types of this association also have a wide compositional range, but are predominantly of monzogranitic composition, in a post-orogenic or post-collisional setting. Both of the I–type classes are associated with copper–base metal mineralisation which, however, is far more strongly developed in the Cordilleran situation. Pitcher, following a suggestion by White, also proposed the recognition of M–types, i.e. granites derived directly from the mantle in oceanic or island arc settings and emphasised the definition of A–types as alkali and peralkali granites associated with anorogenic faulting or rifting, such as the granites of the Nigerian ring complexes, which are tin associated, and also those of the Oslo rift. Thus in this scheme each granite type was assigned to a specific kind of tectonic setting.

There is no doubt that the additional types identified by Pitcher do represent granite populations which have a wide natural occurrence. However, the MISA, or alphabetical scheme, as it has sometimes disparagingly been called, has not proved easy to apply in practice, even though both granite type and tectonic setting are evidently geologically related.

The connection between granite type, source region and tectonic setting has subsequently been addressed by many writers, using criteria as various as the nature of the enclaves (Didier et al., 1982), biotite composition (Nachit et al., 1985), zircon morphology (Pupin, 1980), associated mineralisation (Xu Keqin, 1982), opaque oxides (Ishihara, 1977), major element geochemistry (Debon & Le Fort, 1983, Batchelor & Bowden, 1985) and trace element geochemistry (Pearce et al., 1984).

Barbarin (1990) has ably summarised all these and concluded that they all reflect the composition of the source region, which he and most others envisaged as

SOURCE	GRANITE TYPE		TECTONIC SETTING	
CRUSTAL	Intrusive Two-Mica Leucogranites	S	COLLISIONAL AND POST COLLISIONAL	OROGENIC GRANITES
	Peraluminous Autocthonous Granites(High K-Low NaCa)	S		
	Peraluminous Intrusive Granites (High K-Low NaCa)	S		
MIXED (Crustal) + (Mantle)	Metaluminous Potassic Calc-Alkaline Granites (High K-Low Ca)	I		
	Metaluminous Calc-Alkaline Granites (Low K -High Ca)	I	SUBDUCTION ZONES	
MANTLE	Island Arc Tholeiitic Granites	M		
	Ocean Ridge Tholeiitic Granites	M	OCEANIC RIFTS	ANOROGENIC GRANITES
	Alkaline and Peralkaline Granites	A	ZONES OF CONTINENTAL RIFTING AND DOMING	

Fig. 5. The relationship between source, typology and tectonic setting, modified after Barbarin 1990.

forming a continuous compositional spectrum from wholly mantle to wholly crustal, with a large intermediate area of mixed crustal–mantle composition. He also concluded that there is a link between the composition of the source region and the tectonic setting in which the granites were generated, and in this he supported the earlier view of Pitcher (1983).

The diagram illustrates the imperfect correspondence of source rock with tectonic setting which has resulted in difficulties in the application of the typological system. Barbarin advocated a more complex notation which in some cases included structural criteria. The earlier notation of Chappell & White (1974) and Pitcher (1983) has been followed here.

Many of the schemes summarised by Barbarin (1990) are broadly similar to the expanded typological system of Chappell & White (1983) and Pitcher (1983). Others, such as the major element systems of Debon & Le Fort (1983) and Batchelor & Bowden (1985), the opaque minerals (Ishihara, 1977) and zircon morphology (Pupin, 1980), are empirical or empirically derived and tend to be used in conjunction with typological nomenclatures.

2.3
Geochemical Tectonic Discriminants

The trace element system most widely used is that of Pearce et al (1984) who divided granites into syn-collisional Syn–COLG, volcanic arc VAG, within plate WPG and oceanic ORG. This scheme is based on the relative abundance of certain trace elements, principally Rb, Nb and Y, which characterise different granites from known tectonic setting, and hence, by analogy with their analysed reference data set, the tectonic setting of subsequently analysed granites. Pearce et al. (1984) however stated that 'the fields on the discriminant diagrams strictly reflect source regions (and melting and crystallisation histories) rather than tectonic setting'. This cautionary note seems to have gone largely unheeded and the system, which is easy to use and which in many cases provides useful insights, has been widely and uncritically applied.

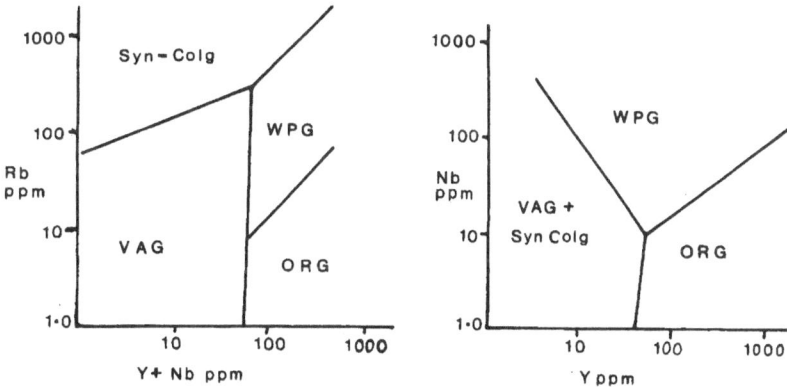

Fig. 6. Trace element discriminant diagram of Pearce et al. (1984) for tectonic setting

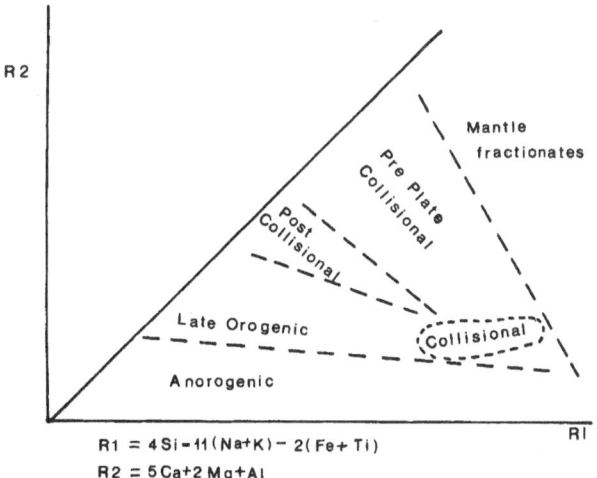

$$R1 = 4 Si - 11 (Na+K) - 2(Fe+Ti)$$
$$R2 = 5 Ca + 2 Mg + Al$$

Fig. 7. The R1–R2 diagram of La Roche (1964) modified by Batchelor & Bowden (1984) to identify tectonic setting

Moreover Pearce and his colleagues were unable to establish a satisfactory geochemical discriminant for post collisional granites which, geologically is probably one of the most important categories.

A major element system for the tectonic classification of granites was developed by Batchelor & Bowden (1985) They adapted the R1–R2 system of La Roche (1978) and utilised it to relate the lineages identified by Bowden et al (1984) to specific tectonic settings. These commonly recurring lineages certainly partially correspond to some tectonic settings. However, the method only works well for granitoids with an extended compositional range. Granite populations dominated by monzogranites, which can occur in a variety of settings, present the greatest difficulty and are not reliably distinguished.

2.4
Comments on the Question of Source Related Nomenclatures

Certain types of granitoids characteristically occur in particular geological situations. Cordilleran I–types (Andinotypes) are almost always found in volcanic arc situations which are most reasonably ascribed to subduction mechanisms of oceanic crust at active plate margins, hence the informal use of such terms as subduction related and volcanic arc granites. Similarly A–type granites are most typical of within plate and anorogenic situations, and sometimes the back arc environment. While typological systems work reasonably well for granites derived from a coherent source like that of the mantle, granites derived from mixed crustal–mantle or wholly crustal sources can occur in a variety of tectonic settings. S–type granites are known in Cordilleran settings (Clarke et al., 1990, Avila Salinas, 1990). They occur mixed together with I–types in settings which are both subduction related and collisional in Southeast Asia (Cobbing et al., 1992) and they occur together in the greater part of the Lachlan Fold Belt.

One difficulty in directly correlating granite type with tectonic setting is that the settings themselves are often very complex, incorporating both mantle and crustal source materials. In the Northern Andes of Ecuador and Colombia dextral strike slip faulting at a transpressional plate margin has resulted in the tectonic juxtaposition of oceanic, volcanic arc and continental slices which host granitoids of I–type character in the two former cases and of S–type in the latter (Litherland et al., 1994). The occurrence of the two granite types within the one belt is entirely compatible with the geology of the region as it is now understood, though at first it gave rise to considerable concern.

Granites may have been affected by many processes during their generation and emplacement but have one thing in common, they have all been derived from some kind of source region. Consequently the identification of this factor provides the foundation for evaluating such processes and for relating the granites to their appropriate tectonic environment. There are of course many grey areas and complete compositional transitions between the main types. Because of these complexities it may be best to regard the MISA and other source related nomenclatures not as a systematic classification, but as a series of signposts within a continuous spectrum of compositional and geological variation

2.5
Additional Nomenclatural Terms

For the most part these are tectonically or structurally based and consequently differ fundamentally in their application to the compositionally based criteria outlined above. Because of this they are inadequate as a basis for classification. Nevertheless there is some degree of overlap or conjunction, and the terms are commonly used by geologists because they appear to be relevant to the numerous particular situations which they encounter in the course of field work. They are briefly summarised as follows:

2.5.1
Tectonic

Convergent
Subduction
    Oceanic Arc   Volcanic Arc   Inner Arc   Back Arc
Collisional
    Syn-Collisional   Post-Collisional
Extensional
    Rift   Uplift Molasse   Within Plate

2.5.2
Orogenic

    Syn-Orogenic   Post-Orogenic   Anorogenic

2.5.3
Structural

    Syn-Tectonic   Post-Tectonic

All these categories are based on regional processes which involve the whole spectrum of geological phenomena. Geologists are quite naturally anxious to establish which of these processes has been operative, and if the granite geology provides some reason for selecting one category rather than another it will inevitably be used. Consequently granites are quite commonly referred to informally as being of Volcanic Arc, Collisional, or Anorogenic affinity and if these affinities are identified on the basis of the overall geological situation these terms can be useful. If, however they are defined solely by the compositional method of Pearce et al. (1984) and without regard to the wider geological context, they may be completely erroneous and lead to the perpetuation of mistaken concepts in the literature which are extremely difficult to correct. In fact it is often very difficult, or even impossible to define satisfactorily the tectonic setting of some Phanerozoic belts older than Jurassic. Consequently these global terms should be used with care.

The more overtly structural nomenclatures usually depend on whether the granite is foliated or not. In fact any granite of any composition or age can be foliated for a variety of reasons. Very often the foliation is synplutonic, related to a specific intrusion mechanism and not to any regional deformation. There are many granites which owe their deformation to emplacement along an active fault, and are quite clearly syntectonic with respect to that structure, examples being the Donegal granite of Caledonian age and the Cordillera Blanca granite of Miocene age (Hutton, 1987, Petford & Atherton, 1992). The wider meaning of syntectonic within an orogenic episode can almost always be shown to be an oversimplification when it is applied to granites.

Because of these difficulties tectonically and structurally based terminologies should be used with care to amplify or support classifications based on compositional criteria.

3 Granite Geology

3.1
The Nomenclature of Plutons and Batholiths

The nomenclature of granite bodies as exemplified by the sequence simple pluton–composite pluton–batholith is apparently straightforward, but in many cases is actually rather difficult. Nevertheless certain ground rules can be established which cover most eventualities.

The fundamental unit of granite geology is the pluton. These are circumscribed bodies which are round, oval, lenticular or tabular in shape at outcrop. If a pluton is composed of a single lithological unit, such as a tonalite, granodiorite or monzogranite, it is termed a simple pluton. If it is composed of more than one lithology it is said to be composite, and it is most likely that the component lithologies will be related and will form an emplacement and differentiation sequence. It is quite common for such plutons to be zoned, usually with the most mafic component at the margin and the most felsic in the core, which is termed normal zoning. In some cases the opposite is found which is termed reverse zoning.

When plutons occur as isolated bodies the procedure is to name the pluton, identify the lithologies and classify it as simple or composite. When a number of plutons are joined together they form a mosaic of interfering bodies termed a batholith. When this is made up of a number of unrelated plutons it is termed a composite batholith. However, in practice the term batholith is generally used in a non-systematic way for any sizeable body of granite whether it is made up of unrelated plutons or not.

3.2
The Anatomy of Batholiths

The actual number of combinations into which plutons can be assembled in batholiths is extremely varied, but the differences can be illustrated by examples of the most contrasted types.

The simplest kind of batholith is made up of simple plutons, such as the Main Range Batholith of Peninsular Malaysia, which is composed chiefly of simple plutons of monzogranite. The most complex batholiths are those of North and South America, which consist of hundreds of plutons which are both simple and composite.

J. Cobbing: LNES 96, pp. 17–65, 2000.

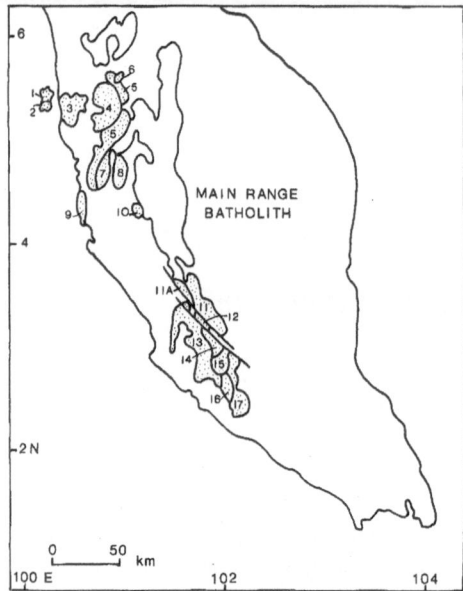

Fig. 8. The Main Range Batholith of Peninsular Malaysia. Most of the Batholith remains unmapped in detail. Mapped plutons are distinguished numerically. 1. Penang. 2. Kampong Batak. 3. Kulim. 4. Selama. 5. Taiping. 6. Bukit Damar. 7. Bubu. 8. Kledang. 9. Dindings. 10. Bujang Melaka. 11. Bukit Tinggi. 11a. Kalumpang. 12. Genting Sempah. 13 Kuala Lumpur. 14. Kuala Kelawan. 15. Jelebu. 16. Chembong. 17. Tampin

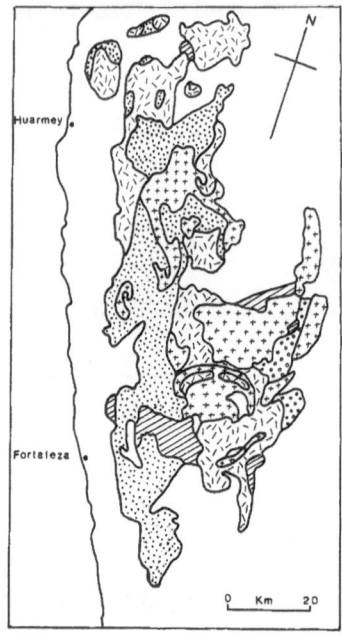

⊡ Gabbro
⊡ Santa Rosa Quartz Diorite - Tonalite
⊡ Santa Rosa Tonalite - Granodiorite
⊡ Paccho Quartz Diorite
⊡ Later Monzogranites
☐ Country rocks and roof pendants

Fig. 9. Simplified map of part of the Lima segment of the Peruvian Coastal Batholith showing the distribution of zoned plutons, super-units and a ring complex

These two directly contrasted types of batholith reflect major geological differences between the granitoids which comprise them, and so are an important expression of actual granite geology. There are also intermediate degrees of combination of simple and composite plutons with a limited or extended range of lithologies, which reflect a different geological status. Examples of these occur in the Caledonides of Britain and Norway, the Lachlan Belt of Southeast Australia and in the Eastern Province of the Southeast Asian Tin belt.

The use of the term complex is usually restricted to any occurrence of granitoids which, because of internal complexity or the intimate association with other igneous or non igneous rocks, are not readily distinguished as plutons or batholiths.

3.3
The Structure of Plutons

Plutons are structurally defined bodies which come in many different shapes and sizes. They may be simple or composite, isolated or mixed with other plutons to form a batholith. Simple plutons are generally round or elliptical in form and tend to have a width of up to about 15 km, though some elongate plutons can have a length of over 100 km as in the Mae Sariang pluton of northern Thailand. In all case however, they consist of a single granite unit with textural or compositional variants which are recognisably derived from it. Composite plutons are generally larger and may be very large, though few exceed 100 km in their greatest dimension. Like simple plutons they are round, oval or linear and are commonly zoned from a margin which may be dioritic or tonalitic, to an inner core of granodiorite or monzogranite, most often forming a genetic rock suite.

Ring complexes are distinguished from other plutons by the presence of a ring dyke which forms a partial or complete rim around an earlier central pluton, which it intrudes. The ring dyke is normally porphyritic with a fine-grained matrix containing megacrysts of feldspars, quartz and dark minerals, the species of which depend upon which kind of granite is forming the complex. In many cases they are full of angular and net veined mafic enclaves in every stage of breakdown into smaller particles. Ring complexes are not normal components of orogenic calc-alkaline plutonism, but they do occur in certain instances, as for example in the Lima segment of the Peruvian Batholith (Cobbing & Pitcher, 1972, Bussell, 1985). They are more abundant in volcanic centres, as in the Tertiary centres of Scotland and Ireland. Most typically however, they form chains of plutons of alkaline affinity in anorogenic regions within cratons, as in Nigeria where such complexes developed successively along crustal lineaments. These alkali granites are commonly mineralised with tin and rare earths.

3.4
Granite Units, Super Units and Suites

In simple plutons the characteristic granite of that pluton is named as a granite unit. In practice the name of the pluton can often be given to the granite unit, so that for example, the terms Belinyu pluton, Belinyu Granite Unit and Belinyu Granite are completely synonymous and intelligible. If however, a granite of identical lithology and texture should occur within two or more plutons, then the plutons should be individually named, and the granite unit separately distinguished and shown as occurring in distinct plutons e.g. the Belinyu Unit of the Belinyu and Penangas Plutons.

The concept of super-units is simply an extension of this principle and recognises that all the lithogical units present within a composite pluton are present within other composite plutons, generally along a plutonic lineament. (Cobbing et al., 1977).

The question of Suites and Super-units is critical for the mapping of granite belts because these phenomena, where present, are initially identified by field mapping. The concept seems to be valid because it was developed independently and simultaneously by geologists working in completely different areas, and has subsequently been found to be applicable in other belts. It is quite likely that it may come to be usefully applied in other belts where it has not previously been recognised. For example, within the Scottish and Irish Caledonides the granites of Donegal, of Connemara, of Leinster, and the Newer Granites of the Grampians, especially Aberdeenshire, form coherent suites of plutons which are evidently locally related, but regionally distinctive. However, a certain caution is required in implementing this concept, because it does not seem to be a feature of all belts and in fact, is virtually absent in some, as for example in Southeast Asia, where all the plutons are distinct (Cobbing et al., 1992).

Whether granites can be identified as rock suites within related plutons or only as a single, distinctive rock suite within a single plutons, they can still be regarded as imaging their source region and can therefore be viewed as deep crustal probes which bring information from lower crustal levels to the surface, where it can be readily studied and interpreted, provided the necessary technology is available. Granites therefore provide an unparalleled tool for discovering the history and evolution of every crustal segment. Their significance in this respect has only recently become apparent. Within any belt there will be many different suites, or plutons, all of which are capable of providing a different window into this past crustal history.

3.5
Granite Belts and Provinces

The largest identifiable assemblage of related plutons and batholiths is the granite belt or province. These are almost always a feature of orogenic belts and they often relate to a particular orogenic phase, though in some geographically coherent belts, granites of several ages may be present. Although there are often similarities between many granite belts and provinces, they are generally quite distinctive and these differences clearly reflect differences in the orogenic processes which led to their formation and the composition of the source regions from which they were derived.

3.6
Contacts

Granite contacts against the country rocks are usually sharp and easily recognised, and can often be delimited on air photographs and satellite imagery. Internal contacts are more difficult to distinguish, especially in forested terrains. However, in arid terrains a surprising number of internal contacts can be detected on air photographs and imagery. In the field these contacts are generally sharp but contacts between some kinds of variants may be transitional. In deformed plutons the contacts with the envelope can be tectonic, and because of tectonic interleaving of the granite with the envelope, such granites are often misrepresented as migmatites, and may even be misinterpreted as being representative of a Precambrian basement.

3.7
Magmatic Textures

In the main, granites are coarse-grained rocks and this means that their constituent minerals can be identified, and the textures resulting from the interaction of their grain boundaries can be readily observed in the field. The combination of these features frequently results in a specific textural and mineralogical fingerprint for a pluton, which enables it to be distinguished from other plutons. The observation of granite textures therefore lies at the heart of granite mapping and of all granite geology. Essentially textures can be divided into two major classes, those which have arisen by uninterrupted crystallisation from a melt, and where crystal growth in some instances may have continued in the sub-solidus state, and those where the crystallisation sequence has been interrupted, giving rise to disequilibrium textures.

3.7.1
Equilibrium Textures

Equilibrium textures are the most important for mapping purposes and for most procedures of classification and analysis. Since they were formed directly by crystallisation from a magma, granites having such textures may be considered as being representative of that magma, and consequently geochemical and other studies of such rocks are likely to be more definitive than those from other textural variants. For the sake of brevity granites with these textures will henceforward be referred to as primary texture granites, since this is the most useful term for their field distinction.

Primary texture granites can be divided into two main groups, hypidiomorphic and allotriomorphic. Hypidiomorphic textures are characteristic of diorites, tonalites and many granodiorites, and in these rocks the texture is dominated by plagioclase and is really rather similar to that of a dolerite. Plagioclase laths are developed as euhedral crystals often with a poor but definite orientation which forms an interlocking meshwork of tabular crystals. Both early and late formed minerals are located in the interstices of this mesh, and in the more acid rocks quartz and K-feldspar are similarly located, and can be readily identified as the last melt fraction to crystallise. In some granodiorites and tonalites the quartzo–feldspathic fraction can be seen in thin section to resorb and replace the earlier formed plagioclase.

Granodiorites commonly provide a transitional region between hypidiomorphic and allotriomorphic textures, the former dominated by plagioclase, the latter by

Fig. 10. Hypidiomorphic granular texture in photomicrograph of the Santa Rosa Tonalite, Rio Lurin, Peru

└─────┘
1 cm

quartz and K-feldspar. In such rocks the importance of the interstitial quartz and K-feldspar can be seen to increase to the point where they predominate. Usually this material forms an anhedral intergrowth of approximately equigranular crystals, but K-feldspar often begins to form larger irregular pools and incipient megacrysts, with a reasonably well defined outline, but with many included, partially digested crystals of plagioclase. This process can be seen to progress, so that in some granodiorites and all the granites, the texture is dominated by quartz and K-feldspar. The K-feldspar may or may not be present as megacrysts. In spite of the apparently euhedral outline of the megacrysts thin sections always show the grain boundaries to be anhedral in detail. The K-feldspar, whether it is megacrystic or not, contains partially resorbed relics of early formed plagioclase and dark minerals. In some granites K-feldspar megacrysts may contain concentric shells of mafic minerals, usually biotite, which outline growth stages of the crystal. These, in addition to the partially digested plagioclase inclusions, suggest that such crystals are in fact phenocrysts.

The transition from hypidiomorphic to allotriomorphic texture can be observed in batholiths with an extended compositional range such as those of North and South America. In batholiths where monzogranites predominate allotriomorphic textures are the rule.

Although the microscopic textures of granites with hypidiomorphic or allotriomorphic granites are really very similar such that in thin section many characteristic features are universally present, their appearance in the field is often very distinctive, and it is this which provides the basis for their discrimination by field methods. In granitoids with hypidiomorphic textures the most distinctive minerals are the mafics, hornblende and biotite, and these occur in a virtually infinite range of shape, size and combination. Hornblende may be acicular or robust, big or small, euhedral or anhedral, in clusters or not with biotite, which will itself have a great variety of crystal shapes from books or barrels to poikilitic flakes.

The mafic minerals are most readily identified in the field, but the felsic minerals are also quite variable, and are often quite specific in form and colour for a particular pluton. By observing such textural minutiae in the field a textural fingerprint for a particular pluton or group of plutons, can be established and this affinity can then be further investigated by geochemical or other procedures.

In granites with allotriomorphic textures it is the felsic minerals which predominate, although such mafics as are present are readily seen, and may be just as distinctive as in hypidiomorphic granitoids. However it is the size, shape, colour and grain boundary relationships between quartz, plagioclase and K-feldspar which provide the main basis for the field description of these rocks. Although most plutons are quite commonly distinctive, there are several broad textural categories which are of widespread occurrence.

The most obvious distinction is between those with K-feldspar megacrysts or without, and whether the K-feldspar is pink or grey. Quite broad categories can be distinguished on this simple basis, and in addition the size, shape and colour of the megacrysts provide further criteria. Granites can also be grouped into two other textural groups, those in which the megacrysts stand out against a background of finer grain size, which however is still quite coarse, and those in which they are not so prominent because the groundmass is exceedingly coarse. In such rocks it is common to find that the quartz is present as reticulate, discontinuous, interlocking clusters which contain most of the plagioclase and the dark minerals. Textures of this sort characterise many Rapakivi granites, but they are also commonly developed in coarse granites of both I and S–type affinity. These textures are of widespread occurrence, and although their mode of formation is not properly understood, they serve to distinguish plutons individually on a field basis. Similar patterns of textural complexity are developed in non-megacrystic allotriomorphic granites, to the degree that it is relatively uncommon to find plutons which are texturally identical. In primary texture granites the only megacystic mineral present is K-feldspar, though very occasionally plagioclase is megacrystic. Normally the only variation seen is in the abundance of K-feldspar megacrysts, quartz or the dark minerals. There is generally not much variation in the grain size of mineral species, though in some cases the dark minerals, and especially biotite, may increase in size towards the margin.

3.7.2
Disequilibrium Textures

Granites with disequilibrium textures are formed when the crystallisation process is interrupted by some event, such as the loss of pressure, and the partial or complete quenching of the remaining magma. If the event occurs early in the crystallisation sequence the included crystals are generally euhedral single crystals of quartz, K-feldspar, plagioclase and mafic minerals which may be rounded or par-

tially resorbed by the fine-grained matrix. Rocks of this sort are most commonly present as dykes or small stocks, either within the granite or, more commonly, as radial swarms within the aureole. They are generally rather simple texturally and are often referred to as granite porphyries. They tend to be associated with potassic granites of I–type character. Disequilibrium textures of more complex character are formed when the quenching event occurs late in the crystallisation sequence when a crystal framework will already have been formed with residual magma located in the intergranular spaces. Quenching at this stage results in the disruption of a partially, or in some cases, wholly crystalline rock, with the result that a mixture of individual crystals and rock fragments are contained within the finer-grained matrix. These are all more or less corroded or, more commonly are in interaction with the matrix. Individual rock fragments can be distinguished by the presence of internal grain boundaries

Fig 11. Sequence of textural evolution in hand specimens from the Tanjong Pandang pluton of Belitung Island of the Tin Islands, Indonesia, a, coarse allotriomorphic texture, b, coarse two-phase texture with abundant granitic megacrysts and lithic clasts in a relatively sparse fine-grained matrix, c, advanced two-phase texture with a smaller proportion of granitic relics set in a fine-grained matrix.

Fig. 12. Sequence of textural evolution in photomicrographs from the Tanjong Pandang pluton, Belitung, Indonesian Tin Islands, a. allotriomorphic texture with interlocking anhedral grain boundaries, b. two-phase texture with xenocrysts and granite fragments included in a fine-grained matrix, c. a more advanced stage, d. a very advanced stage with fine-grained groundmass

1 cm

identical to those present in the adjacent primary texture granite. In the field they display an extremely heterogeneous appearance. They are usually associated with highly evolved S–type and A–type granites and are distinguished from granite porphyries by the complexity of their textural and field relationships.

These granites are most readily distinguished by their polymineralic megacrysts of rounded quartz, K-feldspar, plagioclase and dark minerals especially biotite, which are enclosed in a finer-grained matrix which seems to have crystallised rapidly by some quenching process, such as the loss of volatiles by venting along a fracture. Such rocks are extremely common in granites, particularly in the most highly evolved varieties, and although the process is essentially simple in concept, it has resulted in a bewildering spectrum of textural variants, most of which can be seen to be directly related to the primary texture granite with which they are associated. Although the process evidently relies upon the presence of a residual liquid, it seems that in many cases the bulk of the pluton was already almost completely crystalline with a well developed touching fabric, and the devolatilisa-

1 cm

1 cm

1 cm

tion resulted in the explosive disintegration and subsequent annealing of an already crystalline rock (Pitfield et al., 1990, Stone, 1974). Such rocks are characterised by the occurrence of crystal aggregates and rock fragments with a touching fabric, in a finer-grained matrix with a quench fabric. These texturally complex rocks have been designated two-phase variants to indicate that there are at least two phases of magmatic crystallisation. The textural sequence allotriomorphic granite–two-phase granite–microgranite also corresponds to a sequence of geochemical evolution (Pitfield et al., 1990) and most two-phase granites are generally more highly evolved than their primary texture precursors, hence the necessity for recognising these rocks in the field and evaluating their use in geochemical and isotopic studies. The contacts of these rocks with their primary texture hosts are often diffuse and transitional, but the more fully developed varieties form sheets and small stocks in marginal and roof situations. In some cases isolated stocks of these rocks may also occur. Two-phase granites, and more especially microgranites, frequently provide the locus for greisenisation and mineralisation. The microscopic textures of these rocks can be extremely complex and may suggest several episodes of grain size reduction and subsequent enlargement (Stone, 1974, Pitfield et al., 1990).

Microgranites and mesogranites seem to be the end members of this process in that they are more generally highly evolved geochemically and the individual megacrysts are often strongly resorbed by the matrix which, in the case of the microgranites is texturally similar to that of the two–phase variants. Mesogranites on the other hand seem to have undergone grain regrowth and the textures are often allotriomorphic, but on a much finer scale than in the primary texture parent. Mesogranites may also contain occasional rounded megacrysts which have resisted resorption. Mesogranites seem to be particularly closely associated with mineralisation.

A further range of disequilibrium phenomena which are particularly associated with I–type granitoids, includes such features as tuffisites, pebble dykes, intrusive breccias and breccia pipes. Although these phenomena are of widespread occurrence, they are not generally developed on the same scale as their analogues in the S and A–type granites and with the exception of the various breccias, are commonly not recognised.

Tuffisites are only developed on a relatively

Fig. 13. Two-phase texture in photomicrograph of the Gorgor pluton, Rio Gorgor, a tributary of the Rio Pativilca, Peru. Crystals and lithic clasts predominantly of plagioclase, set in a fine-grained quartzo–feldspathic matrix

1 cm

small scale of metres or tens of metres in dimensions and are usually linear in form, occurring as vertical or horizontal sheets. They are unlikely to be recognised in areas which do not benefit from excellent exposure. They can be observed to develop along shear zones and mylonite zones in the granite. Myers (1975) considered that the mechanically formed mylonites provided avenues for the entry of residual gas charged magma which progressively exploited the initial fracture, resulting in the development of tuffisites and breccias. These intrusive sheets consist of modally and texturally variable microgranodiorite, crammed with small xenoliths, which are often rounded and sometimes mantled by felsic shells. The crystalline matrix contains megacrysts that show evidence of mechanical and chemical corrosion.

Texturally, tuffisites are similar to two-phase granitoids in that they contain crystal and rock fragments in a fine grained quartzo–feldspathic matrix. Because they result from the break up of tonalites or granodiorites with a hypidiomorphic texture, the crystal and rock fragments are predominantly of plagioclase which often retain the grain boundary relationships of the original granitoid. Quartz megacrysts are generally of subordinate importance.

Pebble dykes are similar but are characterised by the presence of rounded rock fragments, usually of the host granite, which are often coated by a thick shell of a dark polished but powdery substance, which has resulted from the blasting and interaction of the rock fragments with the gas-charged transporting medium. Usually the frag-

Fig. 14. Tuffisite, Cruz de Laya, Rio Lurin Peru. Photomicrograph showing megacrysts of quartz and plagioclase set in a heterogeneous quartzo–feldspathic base

1 cm

Fig. 15. Porphyry stock. Acos Upper, Rio Chancay, Peru. Photomicrograph showing two-phase texture with crystals and lithic fragments set in a fine-grained quartzo–feldspathic matrix

1 cm

ments are of the host granite but some pebble dykes contain fragments of the enclosing country rock which have been intruded and entrained in a convective system allowing the fragments to move both up and down within the entrainment column as shown by Cloos (1941) in his classic study of the Swabian tuff pipes.

Intrusive breccias, explosion breccias and breccia pipes are larger structures which carry angular and rounded fragments of all shapes and sizes, supported in a fine-grained, usually highly siliceous matrix. In most cases the rock fragments are highly altered and are often mineralised with sulphides. There is consequently a natural link in this sequence of events with the formation of some kinds of mineral deposits.

3.8
Layering and Schlieren

In many granites thin bands or streaks of dark minerals are seen which are discontinuous, wispy or distorted with highly complex outcrop patterns. In granites associated with migmatites and high-grade metamorphic rocks, it can often be shown that such features result from the incorporation of mafic material from the country rock host, and its streaking out within the granite along flow lines parallel to the margin, or to other included blocks. Other inclusions such as cognate enclaves, original restite or contempory basic material may also be streaked out in the same way. Such features representing streaked out host rock inclusions or other included material have been collectively named as schlieren (Mehnert, 1968). In certain rare plutons where a roof to the granite has been preserved or can reasonably be inferred, layers of alternating mafic and felsic minerals may be present which show no signs of later distortion. These layers grade from dark fine-grained rocks on one side to felsic coarser-grained rocks on the other (Bateman, 1992). In such cases the layers of mafic and felsic crystals have evidently grown in a passive environment and show every sign of similarity to the rhythmic layering seen in gabbros, which result from gravity differentiation.

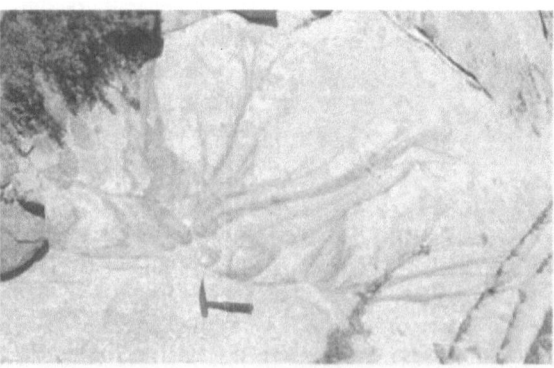

Cumulate layering in granites has been described by (Harry & Emelaus, 1960, Emeleus, 1963) in granites from Greenland and by Townend (1966) and Craxton (1968) from the Omey Island and Galway Granites respectively in

Fig. 16. Schlieren with enclaves Mt Givens granodiorite Sierra Nevada Batholith. Aligned mafic enclaves define a foliation which transects the schlieren bands. Hammer 30 cm

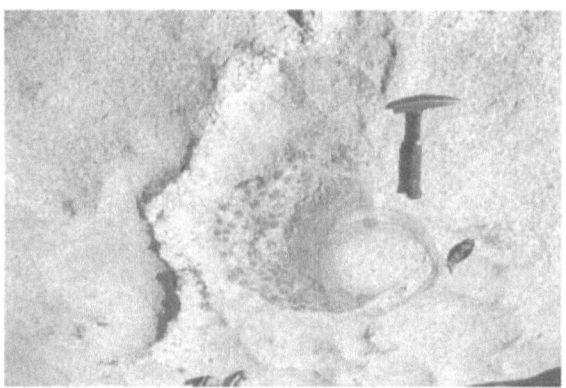

Fig. 17. Schlieren with enclaves, whirlpool structure, Mt Givens granodiorite Sierra Nevada Batholith. Hammer 30 cm

Ireland. These authors have shown that the darker bands commonly contain 30% or more of mafic and accessory minerals such as apatite, zircon and allanite identical in character to those distributed in the main granite, which are poikilitically enclosed in quartz and feldspar. The alkali feldspars in the felsic bands commonly lie with their long axes parallel to the banding. Similar layering in the Dindings pluton of Malaysia (Cobbing et al., 1992) has a similar gradation from mafic to felsic bands, but with the large alkali feldspars in the upper part of the felsic bands having their long axes perpendicular to the banding, and with faceted upper crystal faces mantled by the dark minerals of the succeeding mafic layer. These phenomena indicate a gravitational, or at least a passive environment for their formation. However some authors e.g. Pitcher (1993) do not consider gravity settling to be a viable mechanism because of the high viscosity of granite magmas, and prefer a model of rhythmic nucleation in a stable environment. An alternative mechanism is flow sorting, suggested by Cloos (1936) and investigated experimentally by Bagnold (1954). These and later studies (Barrere, 1976) showed that shear stress in the magma causes grains to collide and disperse in accordance with the squares of their diameter. The effect is to sort the grains by size, the largest grains being displaced into the regions of lowest shear stress (Bateman, 1992).

Although such banding can be recognised in its most unequivocal form in roof zones, it is more commonly present in marginal zones, where it is usually steeply dipping. In some such cases the relation to the type developed in the passive roof is often clear, but more frequently the mafic laminae have

Fig. 18. Schlieren and enclaves within the Mt Givens granodiorite, Sierra Nevada. Hammer 40 cm

Fig. 19. Layering in the Dindings Pluton of Peninsular Malaysia

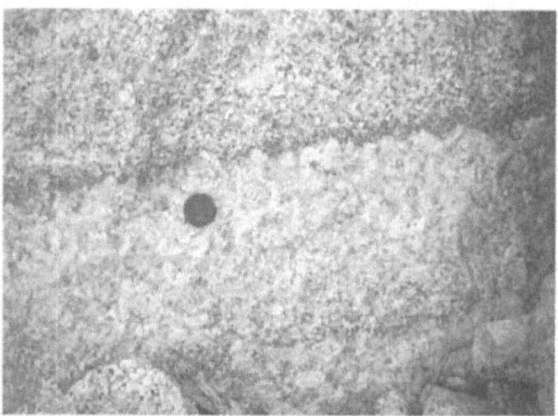

Fig. 20. Layering in the Dindings pluton showing the mantling of the upper crystal surfaces of K-feldspar crystals by an accumulation of biotite crystals. Lens cap 4 cm

been distorted by movement in the marginal zone which results in complex outcrop patterns, some of which resemble cross-bedding and other sedimentary phenomena. In some cases these structures may be intermingled with both enclaves and xenoliths in the marginal zone, suggesting considerable disruption and flowage during emplacement. Although layering such as this and other types (Stephenson, 1990) can be confidently identified as being of magmatic origin, it can become streaked out in exactly the same way as included mafic or metasedimentary material and so can come to resemble schlieren which have quite a different origin.

Poldevaart and Taubeneck (1959) distinguished cumulate layering with tabular crystals parallel to the layering, from comb layering in which the crystals have their long axes normal to the layering planes. Stephenson (1990) identified several types of layering from the upper part of the Hinchinbrook Island pluton of northeast Australia. Compositional layering consisted of adjacent bands of contrasting grain size, which were considered to have formed in response to rhythmic changes in the rate of nucleation. Several kinds of downward facing structures were identified which consisted of quartz, feldspar and amphibole crystals with faceted lower crystal surfaces, and feathery intergrowths of quartz and feldspar hanging downwards from a consolidated upper surface. Stephenson (1990) also distinguished a gross layering consisting of three sub-horizontal granite units within which the

Fig. 21. 'Dropstone' of San Jeronimo Granite bending the xenolithic facies of the Puscao Granite. Qda Quintay, Rio Huaura, Peru Batholith

|———————————|
1 m

finer-scale layering is developed.

The Puscao Pluton in the Coastal Batholith of Peru is remarkable for the development of gross layering expressed as three different facies. The uppermost Tumaray facies consists of horizontal sheets of aplitic granites and pegmatites with a thickness of 700 m, the middle Puscao facies consists of the normal granite, and the the lowest xenolithic facies consists of well defined concentrations of strongly flattened mafic xenoliths forming sub-horizontal bands. These xenoliths have been interpreted as having been spalled off from the volcanic rocks which form the roof of the pluton to fall through the magma chamber and accumulate in layers close the floor of the pluton. This interpretation is supported by presence within the xenolithic zone of a huge block of the earlier San Jeronimo granite tens of metres in size, which has fallen into the

Fig. 22. Gross layering in a pluton of the Puscao granite. The upper part consists of sheeted pegmatites, leucogranites and aplites. Rio Supe, Peru

xenolithic facies and has depressed the xenoliths downwards somewhat in the manner of a glacial dropstone (Cobbing & Pitcher, 1972, Pitcher, 1993).

Taylor (1976) made a geochemical study of the pluton based on the sampling of a statistically determined grid within the Puscao facies. He found that the large-scale variations were the result of in situ differentiation by a mechanism of convection driven thermogravitational diffusion. The examples given above suggest, that in spite of the problem of the high viscosity of granite magmas, a variety of gravity controlled phenomena are of widespread occurrence and seem to be a normal expression of granite geology.

3.9
Residual Fluids

As the magma crystallises the remaining liquid becomes progressively more attenuated and felsic, occupying an increasingly smaller volume. It is possible for a granite to become virtually completely solid and yet retain some residual fluids. Aplites and pegmatites, for example most typically occur along joints which must have developed at a late stage in the consolidation of the granite. However, there are many other manifestations of the ubiquity of residual fluids. They seem to concentrate preferentially at any place where there is a compositional or ductility contrast. Thus it is quite

Fig. 23. Coarse K-feldspar megacrystic granite intruding earlier gabbro. The granite exploited a joint in the gabbro. However the entrance to the fissure was blocked by a large megacryst. Residual fluids percolated around the obstruction into the crack where they crystallised as fine-grained granite. Note also the development of quartz ocelli in the gabbro. Lisa Aragabo pluton Kola Peninsula, Russia. Lens cap 4 cm

common to find a felsic margin developed along a contact, whether it be with the country rock or an internal contact. In granites where mafic dykes are present, and have been synplutonically deformed, the dyke margins are often coated with a film of felsic material which may penetrate as veinlets into the dyke interior. In cases where such dykes are boudinaged the spaces between the boudins are commonly filled with such fine or medium-grained felsic material. Mafic enclaves and other inclusions are often rimmed with felsic material, which seems to react with the enclaves to produce a coarser darker rim. Sometimes these felsic rims seem to react more strongly with the enclaves resulting in their disagregation and dispersion within the host granite as indistinct mafic clots and diffuse darker areas of hybrid material.

The ubiquity of such phenomena suggests that residual fluids may be present and are widely distributed within granites until a very late stage of their cooling history. It is thus quite easy to envisage the possibility that sudden changes in pressure, such as that produced by the action of a fault on a rising granite, could result in decompression and the production of two-phase granites as described above. Granites with a higher level of volatiles are likely to be prone to such activity to a much later stage in their cooling history than other, drier granites. Consequently it is not surprising that tin granites, and certain alkali granites with high levels of fluorine and boron, are particularly prone to be characterised by such rocks.

In granites which have been subjected to synplutonic deformation it is often found that felsic veins are particularly abundant. The earliest veins may be concordant with their hosts and are foliated in the same direction, but later veins may become progressively more discordant in the direction of their emplacement and are also more felsic. They are also progressively less strongly deformed until the youngest veins are completely undeformed. This sequence suggests that synplutonic deformation promotes the migration of residual fluids into linear areas of collection which are then intruded as veins and which record a decreasing level of deformation with time.

3.10
Aplites Pegmatites and Granite Veins

By far the most common types of veins are aplites, which are often widely distributed throughout the pluton and are emplaced as thin veins, mainly in granodiorites and monzogranites and following joint planes. They are fine-grained rocks with a sugary texture which commonly contain biotite and less commonly muscovite. Some aplites are completely devoid of mafic minerals.

Pegmatites are actually not very common in granites, but when the do occur they may form complex veins with aplites and with the aplitic component forming a marginal or central zone. Pegmatites in granites tend to be located in the immediate envelope to the granite roof or walls and also within the upper levels of the pluton itself. These differences in mode of occurrence have been related to depth of emplacement, with those occurring in the country rocks related to a higher level of pluton emplacement than those within the granite.

Varlamoff (1972) distinguished ten categories of pegmatites and quartz veins. They are listed below and the typology of the granites with which they are usually associated is indicated.

Type 1   I–type. Microcline, plagioclase, biotite, magnetite
Type 2   I–type. Microcline, plagioclase, biotite, magnetite, quartz
Type 3   I–type. Microcline, plagioclase, Biotite, quartz, black tourmaline
Type 4   I–type. Microcline, plagioclase, biotite, muscovite, black tourmaline
Type 5   I–S–type. Microcline, Quartz, muscovite, beryl, albitisation
Type 6   A–type. Big beryl, amblygonite, spodumene, columbite/tantalite, well zoned with quartz concentrations, albitisation, greisenisation
Type 7   A–type. Partially or completely albitised, quartz, spodumene, muscovite, greisenisation subordinate cassiterite, columbite, tantalite, white beryl
Type 8   S–type. Quartz veins with large microcline, muscovite, cassiterite
Type 9   S–type. Quartz veins with muscovite and cassiterite
Type 10 S–type. Quartz veins with cassiterite, wolframite, scheelite

While most pegmatites occur within the uppermost levels of granite plutons or within the immediately neighbouring envelope, some pegmatites are found at substantial distances from any known pluton, such that actual relationship to any particular granite is suspect. Pegmatites also form an essential component of vein swarms and while their general relationship to the local granitoid plutonism in such cases is unquestionable it is a matter which has not yet been properly studied.

Pegmatites form an essential feature of many Precambrian terrains where they form extensive swarms. In this case they do not appear to be related to any specific phase of coeval granitic plutonism.

3.11
Enclaves in Granitoids

Enclaves have been observed in granites for many years and their obvious difference from the surrounding granite has led to to a rich vocabulary of descriptive terms both from the geologists who studied the rocks and the quarrymen who worked them. The best summary of these terminologies was provided by Barbarin (1991). Only the most commonly occurring terms are considered here. The term enclave by itself is of a general nature and includes all polymineralic aggregates enclosed in granite. Within that context xenoliths are fragments of country rock of whatever nature, which have been included in granite. Surmicaceous enclaves are not commonly developed and are normally confined to S–type granites where they are interpreted as visible restite i.e. those refractory components of pelitic rocks which have resisted melting during the in situ production of granite melts, and which have been subsequently incorporated as dispersed fragments within the mobilised granite. They consist of micaceous lenticles, possibly containing sillimanite, cordierite and garnet.

However, very small surmicaceous enclaves may be widely dispersed in some unusual I–type batholiths such as that of southern Bohemia. Mafic microgranular enclaves are round or oval microdiorite blobs which are commonly widely dispersed in I–type and A–type granites. Cognate enclaves are small tabular or rounded pieces of related or precursor intrusives caught up in the main body. Of these various categories xenoliths and mafic microgranular enclaves are by far the most abundant.

Xenoliths are present in most granites. They consist of inclusions of country rock broken off as rafts or trains which are generally broken down into comminuted fragments of smaller size, i.e. the true accidental xenoliths.Although they are usually restricted to roof and marginal zones they may be widely distributed through the pluton. In this case they can generally be distinguished as xenoliths because of their similarity to the country rocks. However, if the envelope should be volcanic or gabbroic it is difficult or impossible to distinguish these locally derived xenoliths from the widely distributed mafic enclaves with quite a different origin. Xenoliths of appropriate composition of either sedimentary, metasedimentary or igneous origin, are commonly strongly affected by the surrounding granite, resulting in the hybridisation of the granite and the ultimate disappearance of the xenolith leaving a diffuse and ghostly dark patch.

3.12
The Question of Coeval Felsic and Mafic Magmas

At present there is considerable controversy concerning the role of mafic enclaves in the context of the mingling of felsic and mafic magmas. Such mingling has been suggested on the basis of the known occurrence of composite dykes with felsic and mafic components, and of various phenomena of interaction and hybridisation between mafic enclaves and their granitoid hosts, and also at contacts between granites and earlier or coeval gabbros. Current models favour the association of mafic magmas with the generation and emplacement of certain granitoid magmas. Mafic enclaves, together with various categories of mafic dykes are interpreted as sequential stages in the model. For this reason it is appropriate to consider all these phenomena together, leaving aside the question of whether they are truly related in the genetic sense, or whether their association is purely fortuitous.

The intimate association of felsic and mafic magmas has been well documented from the Tertiary igneous centres of Scotland for some time (Harker, 1904) where composite dykes containing felsic and mafic components were described. In recent years a great deal of research has been devoted to this question and the main lines of the current argument can be briefly stated as follows.

It has been known for many years that certain classes of granitoids contain swarms of mafic dykes which are confined to the granites or only extend for a short distance beyond them. Roddick and Armstrong (1959) demonstrated that mafic dykes within tonalites and granodiorites on Cortes Island in British Columbia, had been broken up into trails of mafic blocks and inclusions of varying shapes and sizes, which had been converted to amphibolites and hornblende schists. They also showed that there were several generation of intrusion of these dykes with the most recent being the least deformed. Intrusion of the mafic magmas had clearly occurred before the host granite had solidified and it was the action of the granite magma upon the cooling basaltic magma which accounted for the disruption.

It had also been observed that in many cases granitoids emplaced into sedimentary or metasedimentary terrains contained evenly dispersed mafic inclusions which could not be matched by any lithology in the surrounding country rocks. This led to the hypothesis that the mafic inclusions had been incorporated at depth from a mafic magma emplaced coevally with the granite. As summarised by Barbarin (1991) it is envisaged that certain granitoids, chiefly I–types or A–types were invaded by dykes of liquid basalt at the time of their generation which, during the ascent and emplacement of the pluton became broken up and disrupted in the manner of synplutonic dykes but, because of the intense convective forces operating during the early stages of granite emplacement, were broken down into smaller rounded blobs and became evenly distributed throughout the granite. It is considered that mafic magmas continued to be available during the emplacement process, and as the granite solidified, mafic magmas continued to be emplaced as dykes. The earliest dykes are now recognised as trains of mafic enclaves. As the pluton cooled later dykes were less strongly disrupted until finally undeformed dykes with chill margins were emplaced during the latest stages after the final cooling. This process is most fully developed in continental margin I–type batholiths such as those of the American cordilleras, but aspects of it are present in most I–type granitoids.

An alternative hypothesis has been developed by some Australian geologists (White & Chappell, 1991) who have interpreted mafic enclaves in granites of the Lachlan fold belt as being the undissolved remnant or 'restite' of material from the source region during the process of partial melting or palingenesis.

3.13
Mafic Microgranular Enclaves

The question of mafic enclaves has given rise to a voluminous literature (Didier, 1987, Didier & Barbarin, 1991). However their most important feature is that often they can be demonstrated not to be xenoliths because many plutons which contain them are not in contact with volcanic formations, and contain xenoliths of the envelope as well as mafic enclaves. They appear to be a universal feature of some classes of granitoids, especially I–types and some A–types, and the patterns of their occurrence are often similar. They do not, however, normally occur in S–type granites, but if S–type granites intrude mafic rocks xenoliths may be included in marginal zones which resemble mafic enclaves. In granitoids of appropriate composition they occur as round or oval blobs of 5–10 cm in size which are usually evenly distributed through the pluton, but which may also occur in trains or in concentrations along internal contacts. They are generally more abundant in tonalites and granodiorites than in monzogranites.

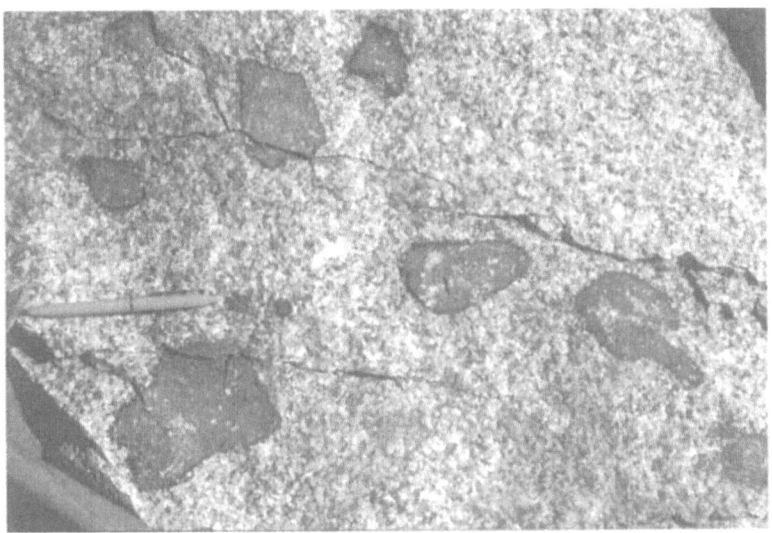

Fig. 24. Mafic enclaves with dark margins. Jerong pluton, Eastern Province Peninsular Malaysia. Pen 12 cm

In many granitoids the mafic enclaves contain megacrysts which are apparently identical to those of their hosts.The commonest megacrysts are plagioclase, biotite and hornblende, which, in their habit are identical to those of the same minerals in the host rock. Thus, if the biotite and hornblende are present as single euhedral crystals in the host, the same will be the case for those in the enclaves. Features such as these are useful mapping criteria. Quartz ocelli are rarely present in enclaves but are commonly referred to in the literature. They consist of a group of quartz crystals mantled by dark minerals, in fact they typically occur in marginal diorites and gabbros, or inclusions of these rocks within the granite. Angus (1962) has described the progressive stages of their formation which he attributed to the hybridisation of older mafic intrusives by later acid material. K-feldspar megacysts within enclaves generally only occur in those granitoids which themselves contain K-feldspar megacrysts, and consequently they are relatively uncommon and are essentially a feature of high K calc-alkaline I–types. In some cases they are oval rather than tabular in outline suggesting a process of rounding, and are often mantled by white plagioclase, the so-called rapakivi texture. They have given rise to much debate in the historical development of granite geology. The transformist school advocated a metasomatic process for their origin and for that of their host granite as well. Modern petrographic studies, especially those by Vernon (1984, 1985, 1986) have interpreted them as being phenocrysts in a granitic magma in which liquid blobs of mafic magma were floating, and which became included in the blobs before they had formed a chilled margin against the cooler granitic magma. At present the model of

Fig. 25. Mafic enclaves evenly distributed in granite, Sillon de Talbot, Northern Brittany. Hammer 40 cm

mingling of coeval felsic and mafic magmas is popular, but there are still some geologists who prefer a metasomatic origin for the K-feldspar megacrysts in the enclaves, even though they consider the granites themselves and the enclaves to be of magmatic origin.

Microgranitoid enclaves have been described from the S–type Wilsons Promontory Granite of the Lachlan Fold Belt, where they occur as swarms and pipes (Elburgh & Nicholls, 1995). These were interpreted as resulting from the mingling of the granite magma with an earlier more basic magma. There has been considerable equilibration of the enclaves with the host granite and they are characterised by such features as K-feldspar megacrysts and quartz ocelli. Although clearly of more basic composition than the granite they do not appear to have the chemical and isotopic features which characterise the microgranular enclaves of I–type granitoids.

3.14
Restite

An alternative interpretation of mafic enclaves has been advocated by Chappell & White (1974, 1991) who state "We take the view that most granite magmas result from the partial melting of the crust. A mechanism of partial melting implies that a silicate melt must, at least initially, coexist with residual unmelted material or restite. Debate about the role of restite in granite genesis must revolve around the extent to which that silicate melt is completely removed from its restite at an early stage in the evolution of the magma. It is our contention that all degrees of separation are possible, so that there is a spectrum of restite involvement. At one extreme is the classical situation in which the granites formed from a largely or completely liquid magma. The opposite case is that in which the magmas retained varying amounts of solid source material within a low temperature silicate melt".

Historically the idea of restite has been inextricably linked with the origin of granitic liquid as partial melt in migmatites with the subsequent inclusion of non mobilised 'resisters'. Chappell and White, in their model of the S–types resulting from the partial melting of pelitic metasedimentary rocks and the inclusion of unmelted material within a granite magma, were in the direct line of descent from Sederholm (1907) and Read (1956).

In the Lachlan Fold belt the S–types contain many inclusions which are clearly of metasedimentay origin and which are readily perceived as restite. Both S–types and I–types exist together in this belt, though to some extent they are geographically distinct, and perhaps as a result of this Chappell and White found no difficulty in applying the concept of restite to the I–type rocks as well, which had come from an igneous source region. Whereas in the S–types the restite was clearly of metasedimentary and metamorphic origin, in the I–types it was represented by the small, mafic microgranular enclaves of dioritic to basaltic composition, which are precisely the phenomenon that many modern researchers prefer to interpret as resulting from the mingling of coeval felsic and mafic magmas during emplacement.

Chappell and White however, did not regard the mafic microgranular enclaves as the most important restite component. They considered the bulk of the restite to consist of undigested crystals of plagioclase, some of which, in the process of reaction with the surrounding granite melt, were partially resorbed or melted and were subsequently overgrown by more acid plagioclase, with the original restite remaining as a core of more calcic plagioclase. Chappell and White consider that magmatic suites result from the differential unmixing of restite material and that the linearity of Harker diagrams reflects this process, rather than the alternative, and generally accepted view of fractional crystallisation. Chappell and White do not deny crystal fractionation but they insist that in some granites restite unmixing is more important. Consequently their view of the applicability of the restite model to all granites results in severe theoretical difficulties for granite geologists, most of whom accept that fractional crystallisation is the main factor in granite evolution once a melt has been formed.

Much research, reported in Didier and Barbarin (1991) has shown that the microgranular enclaves are in chemical equilibrium with their hosts and that consequently, whatever their mode of origin. it is impossible to distinguish them on geochemical grounds. It is also impossible to distinguish the enclaves of Cordilleran I–types from their granitoid hosts on the basis of their isotopic features because the source region from which they originate, the mantle or a juvenile basaltic crustal underplate, has the same isotopic characters as the daughter granite. In particular they do not contain older crustal zircons. Consequently enclaves in these rocks may be either the result of magma mingling or of restite origin. However it is from these rocks that the bulk of the evidence for a magma mingling origin involving the dismembering of early synplutonic dykes has ben recognised.

Enclaves in high K calc-alkaline I–types derived from an older crustal igneous source may have had a crustal history which can be recognised isotopically, and they may also contain zircons with relic cores having much older ages, and providing clear evidence of former crustal residence. Such zircons in enclaves have been recognised in the Lachlan Fold Belt (Williams pers. comm.) and in the Old Woman Pluton of Nevada (Miller et al., 1992). These zircons are manifestly of restite origin. However this does not mean that all the enclaves in those granites are restites. The position therefore seems to be that mafic enclaves may be either of a juvenile magma mingling origin or of restite origin, and that there are no satisfactory field criteria by which they can be distinguished. It may be significant that mafic synplutonic dykes do not yet seem to have been reported from the Lachlan Fold Belt.

From the point of view of the field geologist it is best to be realistic and recognise that field observations are unlikely to contribute markedly to the resolution of this debate. The matter is best reserved for detailed geochemical and isotopic studies. However, mafic enclaves have only comparatively recently been recognised as being important or even critical phenomena, and consequently there is a need to broaden the base of the purely empirical observations from different granite belts. The geologist should first of all note the distribution of enclaves within the pluton, whether they are evenly dispersed, concentrated in marginal zones or in some other pattern of distribution. The size and shape of the enclaves should be recorded and the form of their contact with the host granitoid. The minerals of the enclave are generally in apparent equilibrium with those of the host. Enclaves may be surrounded by a felsic rim and in some cases this material can be seen to be in reaction with the enclave producing a hybrid area of dark minerals of similar size and shape to those of the granite. Ghostly darker areas of granite may result from this process It is also very often the case that the enclaves contain megacrysts. These are most commonly of plagioclase, biotite and hornblende. The size, shape and distribution of these within the enclave should all be noted and whether or not they are similar or identical to those minerals in the host granite. Less commonly enclaves may contain quartz ocelli but this is more usually a feature of granite–gabbro contacts. Enclaves in granites with K-feldspar megacrysts may contain similar or identical megacrysts. These are often located near to, or crossing the margin, but they may also be evenly distributed. Sometimes such megacrysts are of rounded or oval form and may be mantled with white plagioclase. Because of the controversial interpretations of these phenomena it is important that the observation and recording of their field relationships be as careful and as comprehensive as possible.

3.15
High and Low Level Interaction of Felsic and Mafic Material

Many of the features noted above as pertaining to mafic microgranular enclaves are also to be observed at granite contacts, especially roof contacts with earlier basic rocks. The basic material may be simply a precursor and hence essentially comagmatic, or it may be considerably older. If the associated basic material is comagmatic it is permissible to suppose that the processes of interaction of basic and acid material may be analogous to those affecting mafic enclaves. If the associated gabbros are not comagmatic the same inference cannot be drawn.

The Craigballyharky Complex of Co. Tyrone in Northern Ireland Angus (1962) consists of two granite plutons intruded into the Basic Plutonic Complex of Tyrone (Wilson, 1972). This latter complex comprises layered cumulate gabbros together with earlier and later dolerites and gabbros, many of them uralitised, and with an outcrop area of more than 300 square kilometres. The basic rocks of this complex provide the host rocks to these plutons and to several other granite bodies. They form the roof and wall rocks of the Craigballyharky Complex and great rafts of these gabbros have foundered within the granite, together with associated pillow lavas and other volcanics of Ordovician age (Cobbing et al., 1965). Angus described the progressive development of quartz ocelli rimmed with hornblende in gabbros forming the wall and roof rocks, through a number of intermediate hybrid types which, in some cases became mobilised and intrusive. There can be no doubt that in this case the gabbros predated the granites and were not related to them in any way. In fact Hutton et al. (1985) have suggested that they are of ophiolitic affinity.

This example from Ireland is unusual because it is clear that the gabbros which were hybridised by the granite are demonstrably older. The phenomena of hybridisation described by Angus (1962) are identical to those observed in basic rocks of comagmatic association and consequently casts doubt on the interpretation that the phenomena necessarily result from the mingling of coexisting acid and basic magmas rather than by hybridisation of earlier basic rocks by later acid intrusives.

Gabbros and diorites associated with Cordilleran and high K calc-alkaline I–types are in the nature of precursors emplaced in situations marginal to the granites and were subsequently invaded by them. They can to some extent be regarded as coeval and comagmatic, in the sense that although older and fully crystalline, they may still have been pretty hot and able to interact with the later granite magmas in some way. In this twilight zone all the phenomena which characterise the magma mingling situation are thought to develop by the interaction of granite magmas and precursor gabbros in a solid but hot condition. Whether the observed phenomena uniquely define a magma mingling process, or whether they reflect the hybridisation of earlier basic rocks by later acid intrusions, is a matter which at present remains unresolved.

Fig. 26. Granite veining norite with development of pillows, Tregastel. Ploumanac'h Pluton, Brittany. Hammer 40 cm

These phenomena were described by Thomas and Campbell Smith (1931) at Tregastel in the Plouman'ach pluton of northern Brittany, in their investigation of the contact of K-feldspar megacrystic granite with a roof of noritic composition. This locality is characterised by the great diversity of hybrid rocks resulting from the interaction of the granite with the earlier norite. Most of these hybrids contain quartz ocelli and K-feldspar megacrysts and in some of them the megacrysts are rounded and mantled with a margin of white oligoclase. This feature was attributed to the basification of the growth medium provided by the hybrid and it can certainly be precisely matched by some mafic enclaves which do not appear to have been related to an earlier basic intrusion. However, at Tregastel Thomas and Campbell-Smith described the net veining of the norite by the granite with the formation of pillows containing evenly dispersed quartz ocelli and K-feldspar megacrysts studding the marginal zones, which they attributed to the mechanical insertion of the megacrysts into the softened and plastic outer zone of the pillows. They observed all stages of the separation of adjacent pillows by granite but a steady transition to enclave swarms is not developed.

Similar occurrences are known to the writer from Peru, Norway, Russia and Ireland and it seems that many of the phenomena said to be characteristic of the interaction of coeval mafic and felsic magmas are similar to those produced by the interaction of marginal basic rocks with later granites at the level of granite emplacement.

The question of inter-

Fig. 27. Dyke of mobilised hybrid with K-feldspar megacrysts cutting norite with quartz ocelli. Tregastel, Ploumanac'h Pluton, Brittany. Hammer 40 cm

action of granitic and basic magmas at deeper levels is much more difficult to investigate because it occurs beyond the realm of normal geological investigation. The roofs of granite plutons are commonly observed but their roots practically never. There are however some occurrences of mixed rocks which have many of the features mentioned above, which can be shown to have acquired these features at

Fig. 28. Granite veining gabbro with development of quartz ocelli, Bosov,Moldanubian Batholith. Hammer head 11 cm

some deeper level. One such body, the Grimstad Granite was briefly visited by the author in the autumn of 1989.

The Grimstad pluton is of late Proterozoic age and is located on the south coast of Norway. It is a coarse pink K-feldspar megacrystic granite of uniform aspect which is emplaced into granulite facies metasedimentary gneisses. Precursor monzodiorites are located discontinuously at the margins and in a large coastal outcrop in the central part of the pluton which formed the early roof to the pluton. These monzodiorites are traversed by numerous dykes of aplite, pegmatite and the normal granite, and in some cases K-feldspar megacrysts with plagioclase rims are developed in the monzodiorite forming a halo near the dyke margins However, there are at least two distinct monzodiorite units present, with intrusive contacts defining a sequence of intrusion. Both of these units contain abundant dispersed quartz ocelli and less numerous K-feldspar megacrysts. These units are very homogeneous and it is evident that they acquired the characteristics of their distinctive lithologies at some deeper level where they must have been in interaction with the granite magma. Just how deep this zone was cannot be known at present, but there can be little doubt that it was at a deeper level than at Plouman'ach and the other granites mentioned above.

3.16
Dark Margins at Igneous Contacts in Granites

Dark margins are not normally a feature of mafic enclaves but are most typically developed at contacts between granites and associated gabbros and diorites. They have been interpreted by many workers as indicating the coexistence of felsic and mafic magmas with the hotter mafic magma chilling against the cooler felsic (Blake et al., 1965). A contrary view was developed by Bishop (1964) who did not contest the concept of coeval felsic and mafic magmas but considered that chill margins in the plutonic situation could not have survived prolonged exposure to the influence of the crystallising granite. He also showed that dark margins at granite–diorite contacts were generally wider than normal chill margins and he attributed their formation to grain size reduction resulting from recrystallisation and interaction with the adjacent granite and resulting desilication. It may be the case that true chill margins between coeval felsic and mafic magmas are only preserved in the volcanic environment such as the centres of the Tertiary Thulean Province of the British Isles.

3.17
Mafic Dykes

Continuing the theme of coeval felsic and mafic magmas, the basic and intermediate dykes associated with I and A–type granites and especially with the I–type granitoids of the western Americas, are of great interest because of the abundance of their occurrence and the diversity of phenomena displayed. These dyke swarms are generally of the same age as the granite and are known as synplutonic dykes. They are usually of basaltic or andesitic composition. Those which are late in the sequence of intrusion commonly have chilled margins but the earlier ones do not and may even be amphibolitised, and are partially or wholly disrupted by the granite host itself. Evidently at the time of their emplacement the granite was solid enough to fracture yet was still capable of flow. The disruption can take many forms and in some cases the dyke and granite host may be plastically deformed and folded to produce a banded gneiss. However, the most usual kind of disruption is for the dyke to be broken into elongate tabular segments with granite occupying the spaces between them.They resemble box like boudins and although the material filling these spaces may be the normal granite, it is quite common for it to be a leucocratic differentiate of the main granite. This feature of a leucocratic variant along a mechanical junction is of common occurrence. In more extreme cases the dykes may be more completely broken up with the progressive rounding of the smaller pieces, resulting in a linear zone or trail of mafic enclaves

Fig. 29. Mafic synplutonic dyke, Quebrada el Carmen Santa Rosa Pluton Peru

and in such cases the enclosing material will be the normal host granite rather than a felsic variant.

Certain kinds of granites and especially the high K calc-alkaline varieties, tend to be particularly associated with potassic dykes of lamprophyric affinity. These have generally been intruded during the later stages of granite emplacement and are not generally deformed. They are often developed as a radial swarm around the pluton and only to a lesser extent within the pluton itself.

3.18
Emplacement of Plutons

A recurrent theme of granite geology is the mechanism of pluton emplacement which is at least partly because of the apparent difficulty of accommodating large volumes of granite magma within an already solid crust. This space problem as it has become called, was one of the basic problems which contributed to the granite controversy of the magmatists versus the transformists. The latter overcame the problem by granitising preexisting material. However the transformist view, in this extreme form, is no longer widely held and the magmatists who prevailed, were left with the problem. Various mechanisms were called upon to solve this question, and for many years the two favoured models were diapirism and magmatic stoping.

The diapiric model of granite emplacement is based upon the observed fact that diapirs of evaporitic salts can rise through their overburden and be emplaced higher up in the crust in forms analogous to those of granite plutons. Grout (1945) conducted experiments based on these observations and found that a diapir could form, and rise by bouyancy forces, piercing the crust and finally mushrooming out at higher levels. This model has been popular with geologists for many years, and in those plutons which have a strong foliation in their marginal zones decreasing in intensity inwards to an undeformed core, it seemed to be a reasonable explanation. The concept was strengthened by the work of Ramberg (1970, 1981), whose centrifuge experiments on materials carefully prepared to simulate density contrasts between granites and crustal country rocks, showed that diapirism was a feasible process for granite emplacement. These experimental studies have ensured that the concept of diapirism continues to be regarded as a viable process for granite emplacement.

The concept of magmatic stoping was developed especially in the Cordilleran granites of North America and also for many ring complexes, where it was seen to result in the foundering of the central cauldron of the granite roof. In many Cordilleran and Andean plutons, great blocks of the volcanic envelope can be seen to be prized from the roof of the pluton, and in their descent within the pluton, become progressively disaggregated and dispersed. Only rarely has the accumulation of such fragments near a possible floor been observed, but at least the sinking of material within a granite magma is a demonstrable fact.

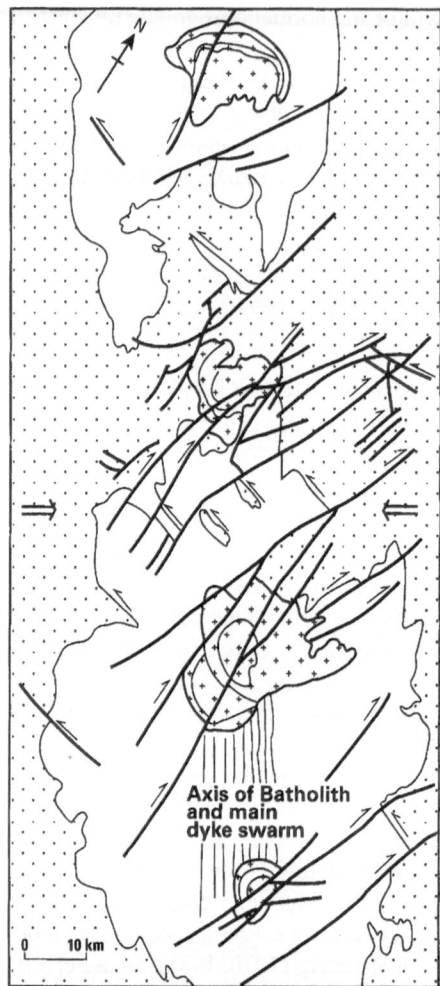

All these intrusion mechanisms require the vertical transfer upward of lighter, hotter granitic material and the downward sinking of the heavier, cooler envelope.

The development of Plate Tectonics has led to the recognition that, in addition to subduction, obduction and rifting, numerous smaller scale but still very large structures are related to the wider tectonic environment and many of these can promote the generation and emplacement of granitoid magmas (Leake, 1990, Hutton & Reavy, 1992). Structures of this sort can penetrate the entire continental crust to a depth where they facilitate the generation of granitoid magmas. Large transcurrent faults in particular can displace whole crustal segments and in doing so can generate a variety of compressional and tensional environments providing favourable sites for granite emplacement.

3.18.1
Stoping

Fig. 30. Structural control of emplacement of the plutons of the Peruvian Coastal Batholith. Vergence of dextral and sinistral fractures indicated

Some of the best examples for illustrating the mechanism of stoping occur within the Peruvian Coastal Batholith. The granitoid plutons were emplaced within an envelope of predominantly marine volcanic rocks which developed within a rift environment at the continental margin (Atherton, 1990). Marine vulcanicity was terminated by uplift and basin inversion, producing elongate shallow dome like structures which provided favourable sites for pluton emplacement by stoping. The south-west–north-east compression which produced these structures also produced conjugate fractures about the strike of the orogenic belt by pure shear (Cobbing et al., 1972, Sylvester, 1988). The rising magmas exploited these fractures and blocks of volcanic country rocks were split off and stoped downwards to be replaced by plutons of dominantly tonalitic composition. The fracture pattern which facilitated this process controlled the pluton boundaries which now have outcrop patterns of tabular or boxlike form with pluton contacts parallel to those of the conjugate fracture system (Cobbing et al., 1972). Intermittent episodes of compression and extension about the main orogenic axis facilitated

Fig. 31. Trains of steeply oriented mafic inclusions resulting from the disruption of a major synplutonic dyke in the Sayan Pluton, Quebrada LLoclla, to the north of the Rio Huauara

pluton emplacement over a lengthy period, controlled the nucleation of ring complexes by fault intersection along the central axis of the batholith, and allowed the emplacement of mafic dykes along the batholithic axis during periods of extension.

The Peruvian example is particularly instructive but similar processes can be identified elsewhere. In particular it may be noted that plutons with tabular or boxlike outcrop patterns may well have been emplaced by similar processes.

It has also been demonstrated by Pitcher and his colleagues that some plutons of typical stock like form which map out as perfectly circular bodies were also emplaced by stoping. The Rosses pluton in Donegal is perfectly circular in plan. However, Pitcher & Berger (1972) were able to demonstrate that in detail the contacts form a jagged margin controlled by the regional joint system, A precisely similar feature was found by Pitcher and Bussell (1977) for the San Miguel pluton of the Peruvian batholith. It is reasonable to infer that many, if not all circular stocks, are characterised by similar features, but in most cases poor exposure does not permit the recognition of such detailed geometry. Stoping is likely to be favoured by brittle rather than ductile deformation.

3.18.2
Diapirism and Ballooning

The concept of diapiric emplacement though once widely held does not now command as much support as it formerly did, and has now been partly replaced by that of ballooning. This historical development is well illustrated by the Ardara pluton from Donegal which was first described by Akaad (1965) as a diapir and subsequently reinterpreted by Holder (1979) as a balloon. In plutons of this sort the shape is generally round, elliptical or drop like, sometimes with a distinct head and tail. The country rocks are normally strongly deformed and structures are concordant with the granite margin. The granite has been emplaced by imposing a regime of ductile deformation on the rocks of the envelope which can provide up to 75% of the required space for the pluton. The Ardara pluton from Donegal again provides the most detailed study of this type of emplacement Holder (1979). During emplacement the metasedimentary country rocks were metamorphosed and deformed under ductile conditions which provided most of the space for the pluton. There were three phases of intrusion. An early outer rim of quartz diorite, an inner ring of coarse K-feldspar megacrystic granodiorite and a core of equigranular granodiorite. Country rock xenoliths of locally derived basic igneous rocks and psammites have been deformed into shapes which reflect the degree of strain affecting the host granite. Those in the strongly deformed outer quartz diorite have been flattened into pancake shapes parallel to the pluton margin. The less strongly deformed inner granodiorites have less well defined foliations, and xenoliths which are only partially flattened or, in the case of the inner granodiorite, unflattened. Holder (1979) attributed the deformation of the country rocks, intrusive components and contained xenoliths to the successive emplacement of granite magmas in the central zone causing expansion of the pluton and deformation of early components of the pluton and the country rocks by processes of ductile deformation.

3.18.3
Patterns of Pluton Emplacement Along Strike Slip Faults

Strike slip faults are ubiquitous in their occurrence and can occur in any part of an orogenic cycle, being essential components of plate movement and readjustment. It has become increasingly clear that some granites of great size are located along ductile shear zones which penetrate the entire crust. It is now thought that these fault zones promote the production of granite melts in the lower crust by adiabatic decompression, and that the magmas rises along the fault zone to be emplaced as plutons along an active fault (Leake, 1990). In such cases the granites may be deformed synplutonically during emplacement and post plutonically after emplacement. They can also be said to be syntectonic with respect to the local structure. The extreme deformation occurring in such cases can result in the development of an orthogneiss. It has now been established that many orthogneisses of this type, which were formerly interpreted as Precambrian basement, are in fact Phanerozoic granites, some of them as young as the Cretaceous or Tertiary.

The Donegal granite is a good example of the history of such a controversy. The early mappers were divided in their opinion. Some regarded it as being Caledonian granite (Kelly, 1853, Haughton, 1862), while others considered the rocks to be of metamorphic or transformist origin (Scott, 1862). Pitcher and his colleagues were able to trace discontinuous metasedimentary horizons of the Late Proterozoic Dalradian envelope through the granite which they termed 'ghost stratigraphy': a concept which at first seemed to substantiate the transformist origin of the granites advocated by the school of Sederholm and Read. Pitcher and his colleagues found that many of the granite–metasedimentary contacts were sharply transgressive, which indicated to them that the granites were of intrusive origin, and that the gneissic fabric in neighbouring granite outcrops was of a synplutonic origin.They described the structure in great detail, and deduced a model of north-west–south-east compression inducing synplutonic deformation on sequential pulses of a plexus of wedge like granite sheets. Later work on the microstructures of the granites by Hutton (1988) showed that the granites had been emplaced along an active sinistral shear zone which had generated a potential lenticular void on a releasing bend. This became filled with successive granite sheets which were transformed into orthogneisses as they cooled. The emplacement of recurrent magma pulses resulted in the isolation of metasedimentary screens of the country rock which, because of the continued deformation, were disrupted into discontinuous fragments which nevertheless retained their original stratigraphic relationship. Both the granite and the metasediments were overprinted by a shear fabric indicating a sinistral sense of movement.

The Donegal region is particularly instructive in that it provides examples of forcefully emplaced plutons along ductile shear zones and passively emplaced granites in the extensional area between the major shear zones. (Hutton, 1982) showed that the main Donegal granite was emplaced by successive injections of magma into a releasing bend of the Leannan fault. However other granites such as those of Thorr, Fanad, Rosses and Toories were emplaced passively as stoped plutons in an extensional area lying between the Leannan Fault and the Great Glen Fault to the north. The Ardara pluton as a diapir, balloon or blister, seems to splay out from the south-west end of the main Donegal Granite in much the same way that Vigneresse (1990) has advocated for plutons associated with the South Armorican Shear Zone. It is thus clear that the interplay of structural features at different scales plays a significant part in affecting the mode of emplacement of granite plutons.

Similar studies have shown that emplacement mechanisms on major strike slip faults are widespread. They have now been documented from

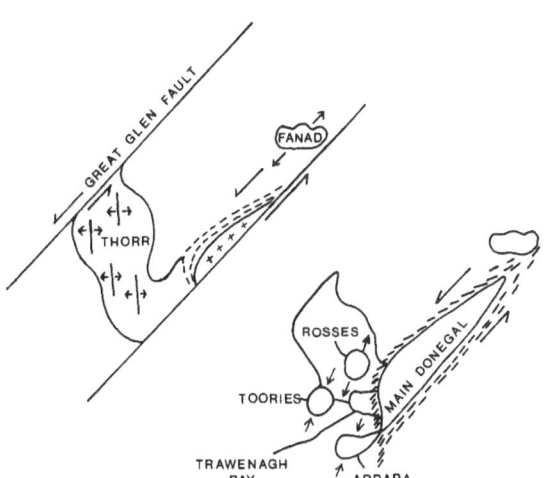

Fig. 32. Structural control of emplacement of a sequence of plutons within a complex strike–slip shear zone, Donegal (after Hutton, 1982)

the South Armorican Shear Zone (Jegouzo, 1980), Central Spain (Castro, 1986), South Korea (Cluzel et al., 1991) and Ecuador (Aspden et al., 1993).

All these studies have depended upon the recognition of shear fabrics in the rocks which indicate the sense of movement of the deformation. The geometry of the microstructures however is equally useful for determining the vergence on thrusts and Hutton & Ingram (1992) were able to demonstrate that the Great Tonalite Sill of Alaska and British Columbia was emplaced along a high angle reverse fault.

3.18.4
Feeder Dykes and Pipes

In recent years the concept of the filling of plutons in the upper crust by narrow feeder dykes has gained increasing acceptance. Pitcher and Berger (1972) identified this process in the Main Donegal Granite, and Hutton & Ingram (1992) have documented an even more remarkable example from Alaska and British Columbia in the Great Tonalite Sill, which is over 800 km in length and consists of innumerable sheets of tonalite, most of them only a few metres thick, emplaced along a high angle reverse fault. Hutton (1992) also documented several instances for the formation of several different kinds of plutons by a process of magmatic wedging and dyke feeding. The sequential emplacement of progressively more felsic material into the central areas of zoned plutons and diapirs or balloons is more readily explained by the presence of a feeder pipe or column beneath the central area of the pluton.

It is probable that the processes of stoping, diapirism, ballooning and dykeing are all effective in granite emplacement, but the present tendency is to pay particular attention to the structural factors operating at the time of emplacement.

3.19
Deformation

Granites are quite commonly foliated during their emplacement by the alignment of tabular and platy crystals parallel to the margin of the pluton. This phenomenon can also be observed in dykes, where such crystals may be aligned in central zones away from the dyke margin. Similar concentrations in the marginal zones of granites suggests that some mineral alignments of this character may be attributed to magmatic flow during emplacement. Another kind of magmatic foliation is produced in plutons emplaced diapirically or by distension. In this case the earliest phases of intrusion become flattened by later increments of magma in the core, which exert an outward pressure on the earlier granite phases and the envelope. This is one way in which an earlier magmatic foliation may be subsequently overprinted by subsolidus, cataclastic and granoblastic deformation. Because it takes place within a pluton during emplacement it is termed synplutonic deformation, and in fact many

Fig. 33. Synplutonic deformation of granite with sillimanite schist. Strong Complex, Peninsular Malaysia. Ruler 15 cm

of the deformation structures seen in granites are of synplutonic character. Deformation undoubtedly begins in the pseudoplastic condition but is most obviously recorded after a crystal framework has formed, when actual cataclasis ensues. Cataclasis involves grain size reduction, and in some rocks where the directional texture is not very apparent, the texture may resemble that of a two-phase magmatic texture produced by a very different process. Indeed, it may even be possible that incipient cataclasis in a rock which is almost fully crystalline, may trigger decompression, volatile loss and the actual production of two–phase granitoids (Pitcher pers. comm.). For the most part however, the solid state condition of the deformation affecting granites is fairly evident.

Deformation in granites normally follows a sequence of progressive changes in the microtexture of the rock. In thin sections the first indication of deformation is the formation of subgrains within quartz crystals or clusters. these may be equigranular with three point junctions, but more commonly the grain boundaries are sutured and extinction is undulose. In most cases the quartz mosaic forms lenticular ribbons of sutured grains, bordered or mixed together with the comminuted or recrystallised relics of magmatic feldspars and mafic minerals. Under appropriate PT conditions these may all recrystallise to give a gneissic fabric to the rock. It is common for the highest strains to be located in particular zones which may have been relatively weak to begin with, and the strong deformational fabric in these zones may anastamose around relatively undeformed areas in which the original magmatic texture can still be identified. The large K-feldspar megacrysts of some granites may often still be identified as relic crystals at quite high degrees of deformation.

Paterson et al. (1989) have outlined criteria for distinguishing magmatic from solid state foliations and although these are multifactorial they consider that magmatic foliations are favoured by the alignment of platy crystals, especially where parallel to external or internal intrusive contacts. They consider that mafic enclaves in ballooning plutons may be flattened by as much as 60% by purely magmatic processes. Magmatic foliations also tend to have a rather homogeneous appearance and distribution. Brun et al. (1990) however, consider that drawing a distinction between magmatic and solid state deformation may be confusing and unnecessary since the tectonic environment for their development may very well be the same, and is in many cases a synplutonic one.

Solid state deformation is often inhomogeneous and is accompanied by recrystallisation and grain size reduction. The strain may be concentrated in particular zones which outline less strongly deformed rock. Fluid activity in zones of ductile shear promotes grain softening and continued deformation.

Some plutons can be quite strongly deformed even when the country rocks are not, and this situation is quite clearly a case of synplutonic deformation. It is also quite common for leucocratic dykes and aplites within a pluton to cut the structures produced by deformation and to be less strongly deformed than the main body, or even undeformed; these record patterns of decreasing strain within a pluton. However, these leucocratic veins are much more common in synplutonically deformed plutons than in undeformed ones which, because they develop complex patterns of folding and refolding, greatly contribute to the complexity of appearance and the difficulty of interpreting these rocks. It is probable that the deformation process itself promotes the segregation of residual interstitial fluids, so that late felsic veins are particularly characteristic of synplutonically deformed granites.

Deformational textures, generally speaking result either from shear or compressional tectonics, though in certain cases extensional tectonics may also result in the development of a fabric. In the case of compression the rock components are rearranged with tabular and platy crystal aligned in a plane of foliation normal to the main compressional direction and with an extensional component in the same plane. This is often seen as a lineation, and in granites this is commonly a mineral

Fig. 34. Leucotonalite orthogneiss in intimate association with amphibolites. The Ban Nong Yai Complex, Thailand. This complex is of Cretaceous age

lineation. The degree of extension and of shortening, may be indicated by strain indicators such as mafic enclaves or the folding of late felsic veins. In many granites there is often a lack of such indicators and it may be difficult or impossible to assess the amount of extension or flattening . In the case of shear tectonics the resulting fabrics are of a rotational character indicating sinistral or dextral vergence in the case of strike slip structures, or upward or downward movement in the case of thrusts, high angle reverse faults, or normal faults.

3.19.1
Field Recognition and Interpretation of Deformational Structures

Foliations are by far the most commonly developed deformational structures and are generally quite readily distinguished. However, the interpretation of the structures is often not very easy, especially in areas of poor exposure. In order to interpret the structures it is first of all necessary to identify the strain markers, or kinematic indicators as they are often called. Many granites are equigranular and lack such features but if a granite has K-feldspar or other megacrysts their shapes will often provide a clue to the dynamics of the deformation process. Similarly if a granite contains mafic enclaves, xenoliths or mafic dykes the pattern of their response to the deformation will be of great help. Quartz is the mineral which is most readily deformed and quartz crystals are first recrystallised into lenticles and then into discontinuous or continuous ribbons in the plane of foliation.

In the case of simple flattening as in a diapir or balloon the foliations are parallel to the margin of the granite and do not show any preferred linear orientation. Enclaves if present are deformed into discoid or pancake shapes within the plane of foliation.

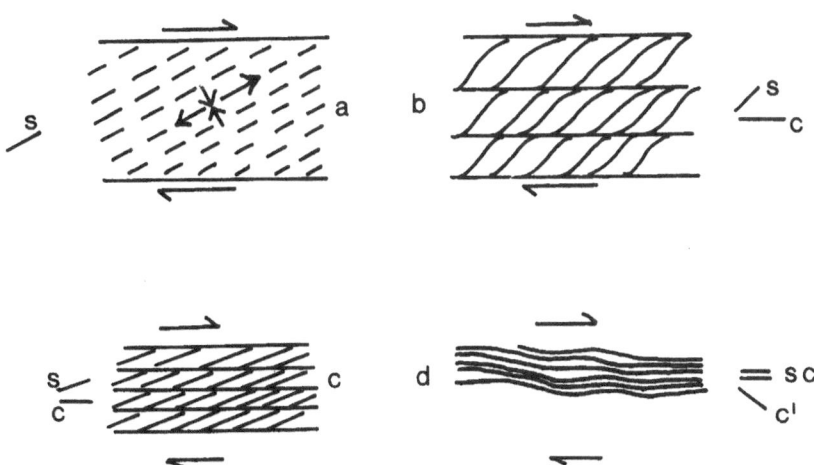

Fig. 35. Kinematic indicators. After Cluzel et al. (1990)

As noted above granites are often generated and emplaced along active strike slip faults and are subject to progressive synplutonic deformation throughout their cooling history. The deformational fabrics formed during this process record the sense of movement operating in the shear zone and in most cases two intersecting planar fabrics are formed. The first, diagonally to the shear planes, is termed an S surface and the second, parallel to the shear planes, is a C surface. These surfaces intersect, either to the right as the outcrop is viewed or to the left. In the first case the sense of movement is dextral, in the second it is sinistral. If the granite is megacrystic the megacrysts will be distorted into a sigmoidal shape indicating a sense of movement either to the right or left as above. Large mica crystals may be similarly distorted into 'mica fish' indicating movement to the right or left in the same way. With increasing intensity of deformation the S and C surfaces converge until they eventually form a single surface which develops into a mylonite. These S–C structures as they are called, are extremely useful and widespread kinematic indicators, but they are not easy to work with and require very careful observation of rather inconspicuous structures. In addition to indicating the sense of movement on wrench faults, they can also be used in the same way for thrusts, normal faults and other kinds of extensional structures.

3.20
Core Complexes

It is only in recent years that a coherent model for the geology of core complexes has been developed (Coney, 1980). It was however prefigured by the identification of 'mantled gneiss domes' from Scandinavia (Eskola, 1948) which in some ways reflect the difficult and apparently contradictory features of the geology of these structures. It is possible that the problematic features of these domes in Scandinavia and Africa (Talbot, 1971) may exhibit some of the aspects of core complexes, though in these cases the formation of the domes by purely tectonic processes cannot be ruled out. The existence of the structures now defined as core complexes has been known for some time in the hinterland of the Cordilleran granites of North America, where geographically restricted complexes of granitoids associated with metamorphic and sedimentary rocks have been recognised for many years, but with no agreed interpretation for their occurrence. However, systematic studies have provided a great body of new data which has enabled the identification of their characteristic features. Most of these have been collected into a special volume (Coney, 1980) in which their defining features have been established. The structures are usually dome shaped with the roof of the dome consisting of sedimentary rocks. These are separated from underlying metamorphic rocks by a detachment zone which is usually mylonitic and is of dome like shape, with kinematic indicators showing outward and downward movement. This detachment surface separates the unmetamorposed cover from underlying metasedimentary rocks in association with deformed granitoids. Isotopic studies on these rocks have shown that the age of the metamorphism is younger than that of the cover.

Since this model has been so carefully and usefully defined, core complexes have been recognised in a number of different tectonic settings, some of which are quite surprising. Within the collisional domain of the Himalayan and European Tethys the Nanga Parbat Massif has recently been interpreted as the result of tectonic exhumation of the gneissic and granitic rocks of the massif by ductile extensional shear, separating the metamorphic and igneous basement from the non-metamorphic cover sequence (Hubbard et al., 1995). Further west in the Menderes Massif of southwest Turkey, movement on a low angle extensional shear zone was accompanied by regional ductile deformation of the basement and the intrusion of two syntectonic granodiorites (Hetzel et al., 1995). Further west, core complexes have been identified in the Cyclades of the Aegean Sea (Lister et al., 1984). It has also been suggested that granite plutonism is an essential factor in the production of core complexes (Lister & Davis, 1993).

Perhaps the most surprising occurrences are on the western rim of the Pacific basin where they have been recognised in the D'Entrecasteaux Islands off the north east coast of New Guinea (Hill et al., 1992). In this occurrence a cover sequence of ultramafic and sedimentary rocks are separated by a detachment zone from a core of mixed gneiss, migmatite and eclogite with undeformed granodiorite. The broader concept of crustal extension associated with domes, antiforms and mineralisation has also been extended to much of the Western Pacific (Mitchell & Carlile, 1994).

Core complexes have also been recognised within the Permo Triassic fold belt of Southeast Asia where, in Northern Thailand a belt of deformed granites and associate metasedimentary rocks with horizontal foliations are flanked by an unmetamorphosed cover (Cobbing et al., 1992). MacDonald et al. (1993) obtained a U-Pb zircon age of 203 ± 4 from the core zone gneisses and a U-Pb monazite age of 72 ± 1 from the same rocks. These results were interpreted as showing a Triassic–Jurassic age for the emplacement of the granite protolith and a Cretaceous age for the deformation. It is very likely that future work in the region of Southeast Asia and elsewhere will lead to the identification of more of these complexes.

Core complexes present considerable problems for geologists engaged in regional mapping programmes and it is unlikely that time or opportunity will be available to fully investigate such a complex, even if one has been provisionally identified. Perhaps the most useful features to be aware of when working with such granitoids is the significance of sub-horizontal or shallow dipping foliations or other structures. The existence of such structures in plutonic rocks can be a useful indicator of the possible role of extensional tectonics.

3.21
Vein Complexes

This is a neglected area in the study of granitic rocks. This is probably because they are usually developed in metasedimentary rocks with a medium to high grade of metamorphism and have consequently generally been perceived as pertaining to the realm of metamorphic rather than igneous processes. Injection Complexes have long been recognised in the Scottish Highlands and were formerly interpreted as having been generated during regional metamorphism as part of the granitisation process. Studies in other regions however have shown that the vein complexes cut their metamorphic host rocks discordantly but have exactly the same deformation structures as some of the discordant granite plutons. Although some recent studies have shown them to be chemically and isotopically related to the associated granite plutons (Singh et al., 1988, Cobbing et al., 1992, MacDonald et al., 1993) it is not yet clear what the relationship is in geological terms. It seems that some vein systems develop in a kind of aureole forming the host to one or more plutons and they seem to be precursors to the main phase of pluton emplacement. However, precursor bodies are generally more basic than their associated granitoids, so the presence of acid precursors is an apparent anomaly. They seem to be commonly present in metamorphic core complexes.

3.22
Migmatites

These rocks have always presented great difficulties of interpretation and nomenclature for geologists because components of evident granitic composition occur mixed together, in an intimate manner with metamorphic rocks of similar or different composition. These difficulties are exemplified by the definitions which have been applied to these rocks.

According to Sederholm (1907) "the constituents of migmatites consist of two elements of different genetic value, one a schistose sediment or foliated eruptive, the other either formed by the resolution of material like the first or by an injection from without". He proposed the name of migmatite (from the Greek word, mixture); the position of this rock group being intermediate between eruptive rocks proper, and crystalline schists of sedimentary or eruptive origin.

According to Mehnert (1968) "a migmatite is a megascopically composite rock consisting of two or more petrographically different parts, one of which is the country rock generally in a more or less metamorphic stage, the other is of a pegmatitic, aplitic, granitic or generally plutonic appearance". Thus the term can be widely used as a concept of purely megascopic structure without referring to a special origin. The term embraces mixed rocks of different origin, e.g. whether the plutonic part is thought to have been intruded (arterite), or whether it has been mobilised in situ from the country rock (venites).

Then according to Winkler (1979) "they consist of composite heterogeneous rocks consisting of preexisting rock material and of granitic material, subsequently intruded or originated in situ. Thus Injection migmatites and in situ migmatite can be distinguished".

It is clear from these definitions that two processes for the presence of granitic material in intimate association with metamorphics exist, a. intrusion of granite magma in the form of veins, and b. segregation in situ by partial melting of granitic material from the metamorphic host. Furthermore, the mafic margins to these segregation are regarded as restite, i.e. a geochemically immobile part of a rock during partial mobilisation of rock components (Mehnert, 1968). These represent two completely different geological processes, one in which a metamorphic host is veined by granitic material to which it is unrelated, the other in which granitic material is mixed with a metamorphic host from which it has been segregated, and is therefore related.

These definitions permit the use of the term for all close associations of granitic and metamorphic rocks and do not necessarily carry any implication for the origin of the granitic material. Many geologists however, continue to use the term in such a way that a metamorphic origin, and probably a Precambrian age for the granitic component is implied, and for this reason the sense in which the term is used should always be clearly stated. It can be readily appreciated that in certain circumstances the use of the term might have been appropriate for some of the examples of granite emplacement, deformation or of core complexes given above.

The distinction between these two kinds of processes is not always easy. However, if the granitic component, whether it it is concordant or discordant to the local structure can be related to adjacent granitoids they are probably arterites. however, if the granitic component is both concordant and is bounded by mafic margins it is probably a segregation granite or partial melt and therefore a true migmatite. it is also possible however that the volumetric increase in the proportion of the granitic segregations may result in the mobilisation and formation of in situ granite bodies of local origin.

The question of the relationship between metasedimentary rocks and anatectic granites has been addressed by several authors Chappell & White (1974) Brown (1973, 1979, 1994) Sawyer (1996). Brown, followed by Sawyer, developed an alternative nomenclature using the terms metatexite and diatexite. For them a metatexite is essentially what other geologists would have termed a stromatic migmatite, that is a high grade metamorphic rock in which granitic leucosomes are bordered concordantly by melanosomes, the body as a whole being characterised by metamorphic structures. A diatexite is a rock in which the granitic component is no longer structurally controlled and has reached such proportions that it

Fig. 36. Segregations of structureless garnetiferous granitoid of S–type composition within pelitic granulite gneiss parent. Toledo, Spain. Hammer 30 cm

becomes mobile, containing disconnected and incoherently oriented relics of the metamorphic precursor. In most cases these consist of banded pelitic remnants and abundant pieces of blue or grey quartz of metamorphic origin. The mobilised granitic component is often heterogeneous containing shadowy remnants of the precursor in various stages of assimilation.

For other writers such a mobilised granitic component has the properties of a magma. Thus Chappell (1996) considered that 'for granite magmas all possibilities exist between pure melts and magmas charged with the maximum amount of solid material consistent with fluid behaviour and corresponding to the critical melt fraction.' Thus Chappell would probably have considered the diatexites of Brown and Sawyer to be magmas with a restite fraction.

These differences of interpretation and nomenclature serve to illustrate the difficulties of interpreting the shadowy area of transition from migmatite to granite.

3.23
Metamorphic Aureoles

Granites which are emplaced into host rocks with a lower temperature, heat up the cover rocks which recrystallise, forming minerals which are stable under the PT conditions of emplacement, which will be different for different levels of emplacement and tectonic control. This phenomenon is known as contact metamorphism and the minerals formed are controlled by the chemical composition of the host rock. Pelitic rocks will recrystallise to biotite schists and gneisses and may contain such minerals as garnet, staurolite, cordierite, andalusite and sillimanite. Impure limestones provide a range of calc–silicate minerals such as grossularite, vesuvianite, epidote and wollastonite. Pure monomineralic rocks such as limestone or quartzite will simply recrystallise without the production of distinctive new minerals.

If the country rocks are volcanics or other kinds of igneous rocks the effects of thermal metamorphism are not so evident, and for the most part rocks of this sort simply recrystallise to a finer grain size, often with an accompaniment of epidote.

Many granites are characterised by the presence of thermal aureoles and since the country rocks are at their hottest close to the granite, minerals stable at high temperature will be present in an inner zone close to the granite, while one or more outer zones may also develop which are characterised by minerals stable at lower temperatures. In favourable circumstances zones of thermal metamorphism can be mapped which help to indicate the temperature of the granite at the time of emplacement.

However, many granites which are undeformed and which were emplaced into lithologies suitable for the production of a thermal aureole, show little or no indication of this feature. Most of the plutons of the Peruvian Coastal Batholith have a very narrow or non-existent development of thermal alteration. In this case the envelope consists mainly of volcanic rocks containing high temperature minerals, and for the most part there is very little sign of a thermal aureole. There are however, areas where the envelope consists of calcareous and pelitic rocks which are intruded by plutons of considerable size. In these situations a very narrow band of recrystallised material may be developed, which is usually not more than 10 m wide and mostly is not even that. This phenomenon has often been attributed to a lack of fluids in the granite. However, the granites of the Main Range Batholith of Peninsular Malaysia do not lack fluids and they too have very narrow or non-existent aureoles.

Some authors, e.g. Chappell & White (1974) have suggested that S–type granites of the Lachlan Fold Belt are characterised by well developed thermal aureoles, whereas the I–types are not. Conversely Hutchison (1977) believes that the I–type granites of the Eastern Belt of Peninsular Malaysia have good thermal aureoles, whereas the S–types of the Main Range Batholith do not. It seems that the development of thermal aureoles is a somewhat capricious phenomenon which does not appear to be related in any systematic way to any particular type of granite or tectonic environment.

Most of the foregoing remarks relate to granites which are relatively undeformed. If granites undergoing synplutonic deformation are in intimate tectonic association with the country rocks of the envelope, both are likely to be deformed together and the local PT environment will resemble that of a regional metamorphic association. This may result in the development of metamorphic rocks with a mineralogy more characteristic of regional metamorphism. Possible examples of this are the Donegal granite (Pitcher & Berger, 1972), and the Stong Complex of Peninsular Malaysia (Singh et al., 1984) where sillimanite garnet gneisses are developed in pelites in intimate association with granite.

The development of metamorphic rocks of regional aspect, but which are actually a result of granite emplacement associated with crustal deformation, has now been well documented from core complexes, and may be more common than has hitherto been considered.

3.24
Mineralisation

Mineral deposits of various kinds are associated with granites and are the result of processes which are an integral part of granite geology. Distinctive groups of mineral deposits are associated with specific kinds of granites, thus porphyry copper and base metal deposits are chiefly associated with Cordilleran and other I–type granites, whereas tin and tungsten are principally associated with S–types and tin and rare earths with A–types.

The processes which result in mineralisation are essentially the same in all granites. Broadly speaking, magmatic differentiation results in the production of progressively more highly evolved granites in which the content of volatiles such as water, fluorine and boron becomes more concentrated. These volatiles are able to carry large quantities of heavy metals. During the final stages of granite solidification and emplacement, magmatic differentiation is superseded by hydrothermal processes. Volatile fluids circulate through the crystalline granite, and in some instances the surrounding envelope, reacting with them and extracting additional metals to those already present. This hydrothermal stage is the natural end product of the geological evolution and differentiation of granites, though only a small proportion of them are ever greatly affected by it.

While the magmatic origin of mineralising fluids can often be demonstrated, it is also the case that a granite may enter the domain of meteoric water during emplacement, which then becomes involved in the hydrothermal process exactly as if it were magmatic water. These different fluids can only be distinguished isotopically. The hydrothermal domain is a complex field of study which currently provides a focus for investigation by the most modern geochemical and isotopic methods.

Indications of hydrothermal alteration in granites are widespread. The most commonly observed feature is the chloritisation of biotite and other dark minerals, and the partial replacement of plagioclase by white mica. Although most commonly observed in thin sections the alteration can also be seen in the field. The formation of films of chlorite and epidote on joint planes is common, and is usually accompanied by the local alteration of the granite. More rarely chloritised joint planes also have a coating of the sulphides of iron, copper or molybdenum.

Granite- related or granite-hosted mineral deposits, tend to have certain features in common, whatever the kind of mineralisation or nature of the granite, but some types of mineralisation are specific to particular granitic suites. The most widely held concept of a mineral deposit is of a vein consisting mainly of quartz, calcite, barytes or some other kind of gangue mineral, and carrying metals in the form of oxides or sulphides. Such veins may occur within the granite or within a limited zone beyond. Disseminated sulphides can occur within porphyritic and brecciated rocks such as porphyry coppers, and oxides such as tin, may be disseminated in leucogranites or massive greisens. Breccias and breccia pipes, either within or outside the granite are also frequently mineralised. Skarn deposits at granite contacts, especially limestones, provide a rich category of mineral deposit.

In tin granites it is common for the hydrothermal processes to be preceded by the production of granites with disequilibrium textures, such as two-phase granites and microgranites. These textural variants are likely to contain fluorite or tourmaline. In the A–type granites of Nigeria cassiterite and columbite are disseminated in the apical parts of the granites as accessory minerals and are the end product of magmatic differentiation.

3.24.1
Endogenous Mineralisation

Mineralisation in granites occurs because residual silica rich liquids collect in the roof zones of plutons. The behaviour of these liquids depends on whether they are impounded or not by an impermeable envelope. If they are, the hydrothermal fluids circulate within the pluton and magmatic evolution is superseded by the hydro-thermal phase. Mineralisation is most likely to occur in or near the roof zone, either in the main body of the pluton, or more often, in the roof of a smaller satellite stock, cupola or projection above the actual roof. In these situations the fluids react with the granite resulting in the production of greisens, which may be either massive or as borders to associated quartz veins. Ore minerals such as cassiterite may be dis-tributed throughout the massive greisens or concentrated in the quartz veins. Such vein swarms are concentrated close to the roof of the pluton and rarely extend downwards for more than a few tens of metres. Neither do they extend far beyond the contact into the country rocks. The best examples of this kind of mineralisation are the tin granites of Southeast Asia and the European Hercynides.

3.24.2
Exogenous Mineralisation

If the envelope rocks are fractured, the hydrothermal fluids are released and mineralised quartz veins develop in the fractures. These veins form at about 500 m beyond the contact and continue for a further 500 m. They are known as fissure veins, and those of Cornish Type are 2 m wide with a down dip extension of 900 m and along strike extension of 2 km. They commonly contain sulphides of copper, iron and zinc in addition to cassiterite and wolframite.

3.24.3
Skarns

The roof and contact zones of granite plutons also provide the main locus for skarn formation in which the country rocks, and sometimes also part of the adjacent granite, are metasomatically converted into massive sulphides and oxides, together with a great variety of exotic mineral species. Carbonates are the most receptive rocks for the formation of skarn deposits but other suitable lithologies include volcanics, pelites and sandstones. Atkin et al. (1985) have shown that the magnetite skarns in the lower Palaeozoic limestones of the Marcona Formation of southern Peru were formed as the result of hydrothermal processes which produced exactly similar mineral parageneses in the fractures of neighbouring Cretaceous granitoid plutons.

3.24.4
Breccias and Breccia Pipes

The granite itself can be brecciated as a result of hydrofracturing and the fractures filled with sulphides and their alteration products. Such pipes are developed within granites, but often continue into the envelope where they incorporate fragments of country rock in the process. Tungsten and antimony mineralisation are present in the tungsten breccia at Khao Soon and the stibnite breccia of Doi Ngom in south-ern and northern Thailand respectively.

3.24.5
Porphyry Copper Deposits

Deposits of this nature are associated with Cordilleran I-type batholiths at continental margins and in Island Arcs, with the former providing a copper–molybdenum association and the latter copper–gold. Although named as porphyries they are actually microbreccias with the sulphides forming a coating to the granitic rock fragments. The porphyries are normally intrusive to the host rock and can be grouped into plutonic, hypabyssal and volcanic types.

1 Plutonic. Formed at depths of between 3 and 5 km. They lack obvious concentric zoning of alteration and mineralisation.
2 Hypabyssal. Formed at depths of 1–3 km. They are most commonly associated with cupolas, plugs, breccias and dykes. Zones of alteration and mineralisation are broadly concentric.
3 Volcanic. Formed at depths of between 0.5 and 1 km. They intrude coeval volcanics in cylindrical plugs or anastamosing dykes and sills. Breccias are common.
4 The host rocks of all these rock sequences are characterised by a multiplicity of intrusive events and textures.

3.24.6
Alteration Zones

Porphyry copper deposits are characterised by zones of alteration which result from hydrothermal activity. Potassic alteration is considered to form from magmatic water only, whereas the phyllic, argillic and propylitic alteration result from the combined action of magmatic and meteoric water. A normal sequence of alteration zones has a potassic core with orthoclase and biotite, a phyllic zone with quartz, sericite and pyrite, an argillic zone of clay minerals and an outer propylitic zone with chlorite.

3.24.7
Epithermal Veins

These veins, which mostly occur in the upper levels of volcano-plutonic arcs, generally have the same regional distribution as porphyry copper deposits. They are a most important source of precious metals and may be associated with base metals. They contain gold and silver in various proportions together with base metals and quartz. They are formed at shallow depths of 1 km–100 m but the downward continuation of the veins carry only base metals. Gold–stibnite veins in Thailand and Burma are not associated with copper porphyryies. Tin is present in epithermal veins in volcanogenic deposits in Bolivia, Mexico and south west USA.

3.24.8
Flow Domes

In Cordilleran situations the latest intrusive manifestations consist of small stocks emplaced at very high levels in the sub-volcanic or even the truly volcanic domain. They provide a wide variety of geological phenomena which are grouped together under the umbrella term of 'flow domes'. These commonly form small circular bodies of stock like appearance and they may be associated with plateau volcanic host rocks. These stock like bodies are often composite and may contain several circular structures defined by a flow foliation within each body. The foliations consist of a compositional flow banding at a scale of 1–5 cm with each band being slightly different, having a greater or smaller proportion of megacrysts and other minerals with respect to adjacent bands. The most prominent megacrysts are square or tabular sanidine crystals. The banding is generally steep in the deeper parts of the structure and is funnel shaped, shallowing upwards and outwards. Hydrothermal alteration is superimposed upon the dome structure. The general sequence of alteration is argillic–quartz alunite–vuggy silica rock–silicified rock with superimposed sulphides. Where fully developed, mineralisation ranges from gold/silver in vuggy silica rock to lead, zinc and copper in the sulphide zone. Many mineral deposits in the South American Andes are of this nature and some of the older mines have recently been reinterpreted in accordance with this model (Cesar Vidal pers. comm.). In Peru the best known example is at Julcani, described by Noble & Silberman (1984).

3.24.9
Field Recognition

The majority of granite associated mineral deposits are located in roof and marginal zones, along contacts and in small outlying stocks and apophyses. Consequently particular attention should be given to these areas during field work, and any indication of alteration or mineralisation of the kinds outlined above should be carefully recorded. In arid regions alteration zones and related phenomena are often quite conspicuous, and may be visible on aerial photographs or satellite imagery, but in regions with significant rainfall and forest cover they are generally inconspicuous or invisible. In areas like these systematic exploration by geochemical and other techniques such as the collection of pan concentrates is the only feasible method of exploration, and it may be difficult or impossible to combine these with regional mapping, though such methods are often routinely employed in sheet mapping programmes.

3.25
Terrains in Relation to Granite Types and Suites

In some orogenic belts granites of a particular type are confined to geologically distinct components of the belt. This is particularly well illustrated in the Southeast

Asian Tin Belt where two granite belts of contrasting type are confined within parallel terrains of different geology. The S–type tin granites lie within a terrain of predominantly Lower Palaeozoic argillites whereas the I–types are confined within a terrain of Upper Palaeozoic sedimentary and volcanic rocks. In this and other belts where convergence of continents and microplates has been a factor it is quite probable that a mosaic of terrains may be present. Barr (1990) has identified three terrains in Cape Breton Island of the Appalachian Caledonides in which the granites within each terrain are distinctive and of different age, with distinct geochemical signatures suggesting a different source region for each terrain. Similarly, recent work in the Ecuadorian Andes has revealed that adjacent belts of highly deformed S and I–type granites are part of tectonically juxtaposed segments of the Andean margin, each of which have undergone a different geological development (Litherland & Aspden, 1995). This pattern of occurrence may well be quite widespread, given the potential for the dispersion and convergence of plates and microplates in a plate tectonic environment. However, such patterns are only likely to be recognised in areas where regional mapping of sufficiently high quality has been done.

The question of terrains within belts and their relation to granite suites is a complex one. Chappell et al. (1988) take the view that granite suites are the defining unit of relationship between granites and their sources and on this basis they have distinguished eleven terrains, which probably could not have been identified in any other way. However, it may be questioned whether these distinct source regions correspond to the more generally accepted view of terrains defined by tectonic criteria.

3.26
Polarity

Geochemical and isotopic polarity have been generally recognised as being related to the subduction process at continental margins, with the granites of the hinterland having a crustal isotopic influence, in contrast to the more primitive signatures near the continental margin. While this may be true for many belts it manifestly cannot be the case for those belts which are built of distinct terrains, which impart their own geochemical and isotopic signatures to the granites derived from them. Alternatively, some belts show geochemical or isotopic polarity along the strike as in Southeast Asia (Pitfield et al., 1987), and also the Peruvian Andes where the Batholith is divided into five batholithic segments.

Granite belts are distinctive and complex geological structures, which only reveal their particular character when viewed in their entirety, or in a sufficiently large area for repetitive patterns to be discerned. These phenomena are simply not perceived at all at the normal scale of geological operations, which is the main reason for attempting to integrate the data at all scales of their natural occurrence.

Such features as granite suites or super-units, the presence of arc polarity, along strike polarity, batholithic segmentation and the distribution patterns of related mineralisation, can only be recognised at the scale of the belt, rather than a portion of it such as a map sheet. This calls for an approach which attempts to tackle the problem at the appropriate scale from the outset. Enough has now been done by geologists to show that this is both possible and necessary, though there will always be difficulties where a belt crosses political boundaries.

4 Characteristics of Granite Typologies

4.1
Origin of the Concept

Many geologists consider that the typological system, first enunciated by Chappell & White (1974) and subsequently elaborated by many others, most persuasively perhaps by Pitcher (1983) and Barbarin (1990), is profoundly flawed, and they have expressed these reservations in fairly trenchant language. Clarke (1992) in his excellent text book on granites, undertook a rebuttal of the concept and defended the idea that classical geological and geochemical parameters characterised granites perfectly adequately. Similarly Atherton (1993) has pointed out that 'the chemistry of the source is often the result of a long history of mantle and crust enrichment/depletion which may have no relation to the tectonic setting prevailing on magma genesis'. This problem of the influence of the source region, either within or independent of, the prevailing tectonic setting, is also inherent within the scheme of Barbarin (1990) and of Cobbing (1990) who considered the role of the tectonic setting to be subordinate to that of the source region in determining the typology of crustal granites.

Critics of the concept consider the specific linking of granites with aspects of plate tectonics to be both premature and misleading, incorporating complexities which might not be appreciated by young, inexperienced and possibly over enthusiastic geologists. Certainly these intimations of caution are well merited, but it is not actually possible to put the genie back in the bottle.

It has to be appreciated that the typological system in its present form provides no more than a framework within which the great variety of granitoids can be loosely accommodated. Its virtue is that different aspects of granite geology can be more readily understood if they can be shown to reflect or characterise different aspects of global tectonics, and this is certainly the case for most granites. This is not to deny that problems remain, some of them severe. We do not know why mid to late Proterozoic orogenic belts are characterised by Rapakivi granites, nor why magnesian-rich durbachites are abundant in the Central European Hercynides. These are but two examples of possibly many exceptions to the consensus model, based as it is principally on models derived from Phanerozoic geology. Nevertheless the underlying appreciation that granites are a reflection of global tectonics, whatever form they might have been, provides a unifying principle which, if carefully applied should lead to elucidation of problems in those areas of uncertainty. It is with these caveats in mind that the the characteristics of the different

J. Cobbing: LNES 96, pp. 67–82, 2000.
© Springer-Verlag Berlin Heidelberg 2000

typologies are outlined, especially with reference to their field occurrence, with the expectation that it 'may help us' to understand the origin and evolution of granitic rocks. In view of the objections to the concept noted above it is perhaps surprising to find that most granites, in the realm of their natural occurrence, are characterised by a range of properties which enables them to be viewed within a typological framework.

Although it has been argued that the typological system was never intended for use in the field investigation of granites (Pitcher, 1993, p.237) it is clear from the publications of Chappell and White that they were first alerted to the possibility of distinguishing granites on this basis by their field experience. They identified a number of field criteria by which S and I–types could be distinguished, which were augmented by geochemical and isotopic criteria. Consequently it is appropriate to approach the question in the same manner, and consider those field properties which help to distinguish these granites. In fact the most typical examples of the various classes are quite readily distinguished by their field characteristics. There are however, numerous intermediate categories which are more difficult. The field properties of the most typical examples are in general fairly straightforward, and although some of them have turned out to be non-specific in character, they allow the field geologist to become attuned to the distinguishing features of granites which, on first inspection, seem to be really quite similar. In some cases a single field observation within the framework of the prevailing background, may turn out to be decisive. This is not often the case however and discretion has to be exercised until more definitive geochemical and isotopic data has been acquired. The following is an attempt to outline the field characteristics of the different typological groups, since this is the basis on which they are first encountered. While there are real differences between the groups there are also real transitions, which may never be properly resolved. Nevertheless the field criteria can provide a first approximation for the distinction of granites in a useful way which, in conjunction with normal analytical procedures, may be confirmed, changed or modified.

4.2
The I and M–Type Family of Granitoids

The I and M–type group of granitoids form an extended family within which several distinct categories can be distinguished, each of which is of widespread distribution, both in time and space. They form a complex group which have given rise to many of the controversies which have bedevilled granite studies. Nevertheless, three main members of this group are far more abundant than the others and provide the main frame of reference for their occurrence and study. These are the M–types of the oceanic island arcs, the I–types of the continental margins, especially those of North and South America, and the I–types of the post-collisional/uplift–molasse association, mainly occurring in continental interiors or in orogenic belts accreted to continents.

Although the M and I–types represent different groups of granitoids within the typological context, there are so many geological similarities that in many cases it is impossible to distinguish them on geological grounds. The reason for this is that granitoids of the oceanic island arcs are produced by exactly the same mechanism as the granitoids of the continental margins, namely, subduction of oceanic crust. The only difference in their geology is that the island arc granitoids are wholly of mantle origin, whereas continental margin granites may have a crustal component, albeit juvenile, and are consequently liable to develop more highly differentiated members. The geological characteristics of these granitoids are so similar that it is not reasonable to consider them separately even though they can be attributed to two different typological groups because of the difference in their source regions. The plagiogranites of the mid oceanic rifts were generated by extensional tectonics and are quite different in character from either of the subduction related groups.

The M–types are divided into two classes. The plagiogranites of the ocean floor, which are only found in continental regions within obducted slices of ophiolite and are of rare occurrence, and the quartz diorites and tonalites of the oceanic island arcs.

Plagiogranites occur as small bodies within the upper parts of cumulate gabbros of ophiolite complexes and are characterised by micrographic textures and the presence of late-stage minerals, suggesting the circulation of a sodium rich hydrous phase. From the point of view of regional mapping such rocks are not likely to be important and will not be further considered. The M–types of the oceanic island arcs have many features in common with the more basic members of the Cordilleran I–types but they are of more primitive character. They are generally considered to have been formed by a mechanism related to the subduction of oceanic crust beneath the island arcs.

Cordilleran I–types are also generally considered to be subduction related at the borders of major continents. They are more highly evolved than M–types and although principally tonalitic and granodioritic they also have a proportion of monzogranites.

Post-collisional, or uplift–molasse-related I–types provide a bimodal association consisting predominantly of monzogranites and granodiorites in association with minor basic intrusives, many of which are appinitic. Swarms of lamprophyre dykes are also characteristic of this association. These granites may also occur in close association with S–types which may be structurally separated, as in the Southeast Asian Tin Belt, or intermingled as in Southeast China, the Lachlan Fold Belt and the Scottish Caledonides.

These three categories of granitoids reflect a progressive increase in the proportion of juvenile, or more ancient crustal material in their generation. The plagiogranites of the ocean floor were derived directly from the mantle and the M–types of the oceanic island arcs were formed during the later stages of Arc formation and thickening, whereby the source rocks may either have been of mantle origin, or of early arc volcanics with a crustal residence time of about 20 Ma Kay et al. (1990). Similar models have been proposed for the Andean Batholith of Peru Atherton (1990), and they may well also apply to other Cordilleran granites.

However, the granitoids of the North American Cordilleras, as exemplified by the Sierra Nevada and Baja California batholiths, are only similar to the Andean batholiths in the western part of their outcrop. Granitoids of the eastern part are characterised by significantly higher isotopic values Kistler & Peterman (1973), and in this inner arc region granites with a crustal signature are present. Pitcher (1993) has suggested that some of these crustal granites such as the Idaho Batholith are similar to the uplift molasse I–types characteristic of granites of the European Caledonides, the Lachlan fold belt and certain granitoids of Southeast China and Southeast Asia. In this latter domain, of granites produced from older continental crust, it is not unusual to find I–type granitoids in the same region as S–types. This clearly shows that the lower crust is a region of complex geology, consisting of older igneous and metasedimentary material of all kinds, and very probably mixed together in all proportions. In view of the complexity of this latter source region it is not surprising that the granites resulting from its mobilisation are sometimes difficult to classify on anything other than an empirical basis. It is with these complexities in mind that the following guide for the recognition of the different types is presented.

4.2.1
M–Types

Occurrence and tectonic setting
Confined to oceanic island arcs and arcs which have been accreted to continents.

Composition
They have an extended compositional range from gabbro to monzogranite, however they consist predominantly of diorite, tonalites and granodiorites. the more silicic varieties range from 64% to 70% SiO_2.

Character of plutons
These granitoids do not form batholiths. Plutons are both simple and composite and may include units of gabbro, diorite, monzodiorite, tonalite, granodiorite and monzogranite. The more felsic varieties may contain aplites.

Mafic minerals
Ortho- and clinopyroxene, hornblende and biotite.

Texture
Gabbros and mafic diorites are crystal cumulates. More felsic rocks have hypidiomorphic textures with interstitial quartz and K-feldspar.

Mafic enclaves
All inclusions are mafic and similar in composition to dykes and country rocks of basaltic and andesitic composition.

Mafic dykes
Dykes are common and some are synplutonic, emplaced while the pluton was still ductile.

Patterns of occurrence
They tend to occur as isolated simple and composite plutons. Cumulate layering is present which is often vertically oriented.

Mineralisation
Porphyry copper deposits with associated gold are present in some eroded island arcs.

Geochemistry
Metaluminous, tholeeitic, calcic to calc-alkaline.

Isotopes
Low initial Sr ratios below 0.704. Similar to Cordilleran granitoids but more primitive in character.

Magnetic susceptibility
High

Sources
Mantle and juvenile mafic crust.

4.2.2
Cordilleran I–Types

Occurrence and tectonic setting
Subduction related settings at continental margins, particularly the Cordilleran ranges of North and South America and, by analogy, with older marginal belts which have been incorporated into continental interiors by subsequent plate tectonic processes. They are typically associated with volcanogenic basins in volcano-plutonic arcs.

Composition
They have an extended compositional range from gabbro to monzogranite, but the dominant lithologies are tonalite and granodiorite.

Character of plutons and batholiths
Plutons are both simple and composite. They are commonly zoned from a basic margin of diorite or tonalite to an acid core of granodiorite or monzogranite. Gabbros are present as large, internally heterogeneous precursor plutons. Plutons of all kinds are mostly coalesced into an interlocking mosaic forming the immense linear, composite batholiths which characterise Cordilleran geology.

Mafic minerals

Biotite, hornblende and pyroxene. The mafic minerals are the most useful indicators for defining granite units because their habit and mode of occurrence tend to be specific for particular granite units. The size and shape of individual crystals, whether they are separated or in association with other mafics, whether they are well defined or poorly defined aggregates, are all useful field criteria.

Textures

The predominant textural mode is of a hypidiomorphic plagioclase mesh, but in some granodiorites there is a transition to allotriomorphic textures, and in monzogranites the textures are wholly allotriomorphic. Equilibrium textures predominate and disequilibrium textures are subordinate, though they they do occur in mineralised plutons and in some potassic suites. Textures are typically medium grained and equigranular. In some granodiorites and monzogranites K-feldspar megacrysts may be present and if so they are usually pink in colour.

Mafic enclaves

These are virtually always present, even in the granites, though they are most abundant in granodiorites, tonalites and diorites. They commonly contain evenly distributed small megacrysts of the characteristic dark minerals of the host, as well as plagioclase. This is one of the features which contributed towards the recognition of super units in the Peruvian Batholith.

Mafic dykes

These are commonly present and very often have chilled margins against the host granitoid. In some cases they have been synplutonically deformed, resulting in the formation of locally extremely complex outcrop patterns.

Patterns of occurrence

Although the geology of a batholith is extremely complex, the geology of individual plutons is often relatively straightforward, with the granites forming good mapping units. Commonly a distinctive granite suite will be present in several closely spaced plutons which may extend in a chain for tens or hundreds of kilometres. These units and super-units enable the gross geology of the batholith to be established. In addition Cordilleran batholiths may be segmented, That is, a specific batholithic segment, usually hundreds of kilometres in length, is characterised by a particular set of super-units, while those on either side will have a different set. In this way there is a variation in composition along the length of the batholith, and it is common to find that one segment is mineralised while an adjacent one is not. In some batholiths there is a compositional variation across the strike of the batholith and this is often accompanied by a rise in the proportion of certain isotopes towards the continental interior e.g., Baja California (Silver & Early, 1977). This phenomenon is known as arc polarity.

Mineralisation
The characteristic mineralisation is of base metal and porphyry copper deposits.

Geochemistry
Generally metaluminous and calc alkaline. Regional lineages within this group may be calcic, sodic or potassic. Some batholiths such as the Cordillera Blanca in Peru are peraluminous with a muscovitic facies, though other subordinate facies, are hornblendic (Atherton & Sanderson, 1987).

Isotopes
The values for $^{87}Sr^{86}Sr$ are generally low though they can rise towards the back arc region.

Magnetic susceptibility
Generally high values. A magnetite series. However some plutons have low or erratic values and isotopic data obtained from such plutons may prove to be unreliable.

Sources
Because granites only occur in any volume in the continental crust geologists formerly considered the crust to have been instrumental in their formation. However, the isotopic values for Cordilleran granites are very low, which suggests either a direct origin from the mantle or some kind of two stage process whereby the crust is underplated by juvenile basaltic magma which subsequently provides the source region for the granites.

4.2.3
Post Collisional/Uplift–Molasse Associated I–Types

Occurrence and tectonic setting
The most characteristic settings for this group are post-collisional or post-orogenic, uplift and molasse associated. They have also been identified in tensional, rift and subduction related settings. They are of widespread occurrence, especially in the Caledonides of Europe and North America, the Lachlan fold belt of Australia, the Eastern Cordillera of Peru and the Eastern Province of Southeast Asia. They are likely to occur in association with S–types.

It is possible that the tectonic setting for this association may be one of tectonic readjustment following an orogeny. Several authors have suggested that orogenic thickening is unstable, and that a form of tectonic erosion may be brought into play by delamination of the lower crust and upper mantle. In this scenario it is envisaged that the orogenic root, together with the underlying lithospheric mantle, breaks off and descends into the non-lithospheric mantle. This creates a post-orogenic situation in which an extensional rift–molasse volcanic association is developed, together with associated granitoids and leads to the preservation of long-lived erosion surfaces, often strongly concordant with the roofs of the associated granites. It corresponds to a development of the older concept of epeirogeny. If true this hypothesis could account for the thin crust and relatively high altitude of the North American Basin and Range Province as well as other regions such as Southeast Australia and Southeast China (Pei Rongfu & Hong Dawei, 1995).

Composition
Like Cordilleran I–types they have an extended compositional range from gabbro to monzogranite, but unlike them the predominant lithologies are monzogranites and granodiorites, with a smaller proportion of tonalites, diorites and gabbros. Consequently the distribution of lithologies tends to be bimodal. A particular feature of this association is the presence of small bodies of coarse hornblendic diorites and gabbros, often characterised by breccias and other features indicative of a high volatile content. These rocks have been designated as appinites and Pitcher (1993), has drawn attention to their petrogenetic significance. He considers them to form a distinctive precursor event and that plutons and batholiths often occur within crude haloes of small, inconspicuous appinite bodies.

Character of plutons and batholiths
Plutons are both simple and composite with many of the latter being zoned. However, simple plutons tend to predominate. Batholiths are composite and tend to be small and linear, in marked contrast to the huge batholiths of the Cordilleran I–types. The also tend to be more scattered in their distribution.

Mafic minerals
Biotite, hornblende and pyroxene. Some monzogranite plutons may only carry biotite.

Textures
Allotriomorphic textures predominate but hypidiomorphic textures characterise the more basic lithologies. The grain size is from medium to coarse with coarser granites predominating, and the texture may be equigranular, inequigranular or megacrystic. If present K–feldspar megacrysts are generally pink but may also be white or grey. Equilibrium textures predominate, but disequilibrium textures are more common than in Cordilleran I–types and are often present in the more highly evolved members of a differentiation sequence.

Mafic enclaves
These are generally present but are not so abundant as in the Cordilleran I–types.

Mafic dykes
These may be present but are not so common as in Cordilleran I–types. However lamprophyre dyke swarms are commonly associated with these granites.

Patterns of occurence

Plutons tend to occur in small groups, often associated with small linear batholiths and forming distinctive sub-areas within major belts. The granites provide good mapping units and in some belts each pluton is texturally unique, but occasionally a distinctive granite unit may occur within several plutons. In other belts distinctive suites of granite units may be present as in the Lachlan Fold Belt. Compositional variation along strike is shown by the presence of distinctive sub-areas. In some cases variation across strike, arc polarity, is present but in general it is not so apparent as in Cordilleran granites.

Mineralisation

In general granites of this association are not strongly mineralised. However, occasional porphyry copper deposits are present, although the predominant mineralisation is of base metals. There may also be rare tin occurrences, and tungsten mineralisation is quite common.

Geochemistry

The granites are generally metaluminous and calc-alkaline, and because of their high K content, define a compositionaly distinctive group designated as high K calc-alkaline (Bowden et al., 1984, Roberts & Clemens, 1993). The more highly differentiated granites are commonly slightly peraluminous. Lineages are generally potassic but some are sodic.

Magnetic susceptibility

Values can be both high and low with the granites belonging to either the magnetite or ilmenite series.

Isotopes

The values of $^{87}Sr^{86}Sr$ are usually moderate from 0.707 to 0.710 and suggest a crustal influence.

Source regions

These granites are generally considered to have been derived by the partial or complete melting of older basaltic material forming the lower crust, and which has had a lengthy period of crustal residence. In some cases there is also a distinct mantle component, and these granites are generally considered to result from a mixture of crustal and mantle sources.

4.3
The A–Type association

A–type granitoids are similar in some respects to I–type granites. However, there are many differences, especially in the manner of their field occurrence. Probably the most significant difference is that whereas I–types are virtually always associated with orogenic processes, A–types are for the most part anorogenic. They occur within continents as fault or rift-associated granites, where typically they form chains of ring complexes. A similar anorogenic setting is that of the Tertiary volcano-plutonic centres of the British Isles which, in some cases, are cored by alkali granite ring complexes. The Rapakivi granites of Scandinavia and Russia also form the cores to high level sub-caldera complexes and are located along deep faults, where they form major batholithic complexes. However, although these essentially anorogenic situations are most typical of A–types they do also occur in orogenic situations, most typically perhaps in back arc areas, where they provide the most convincing evidence for arc polarity (Brown et al., 1984). They also occur in late and post orogenic situations. The latter have been recognised in the Lachlan Fold Belt where small plutons of biotite granite which intrude earlier I–types have been found to be of A–type affinity. In this case they are interpreted to have been derived from a source region which had been depleted by the previous extraction of I–type melts (White & Chappell, 1983). Similarly, in Corsica successor A–types intrude Hercynian calc-alkali granites which are 100 Ma older. In most of these cases the association with deep crustal faulting, rifting and fault intersections emphasize the essentially anorogenic and fault-related association for the A–types.

A–type granites also occur however in a completely different environment, the oceanic islands lying on the mid-oceanic ridges such as Reunion, Ascension and Kerguelen. These represent a relatively young pile of oceanic basalts, intruded by clusters of shallow seated volcano plutonic central complexes. Felsitic rocks are subordinate but resemble the granitoids of the Oslo Rift and of Nigeria (Lameyre, 1983). These oceanic islands however are not likely to fall within the remit of most field geologists.

4.3.1
A–Types

Occurrence and tectonic setting
These are rather distinctive, because A–types are not normally constituents of orogenic belts except in back arc situations. They are most typical of anorogenic and rift settings, where they are strongly associated with major faults, and also with alkali and calc–alkali volcanicity.

Composition
They have an extended compositional range from gabbro to syenogranite with a bimodal distribution and with monzogranites, which are often alkalic, predominating.

Mafic minerals
Biotite, hornblende and pyroxene, the latter two are often sodic. Some plutons may only have biotite.

Characteristics of plutons and batholiths

Plutons are both simple and complex and it is granites of this type which most frequently develop ring complexes. Chains of migrating ring complexes are often located along fractures, but these nested groups are not generally considered as batholiths.

Textures

Textures are allotriomorphic in the more acid rocks and hypidiomorphic in the intermediate and basic. The grain size ranges from medium to coarse and is mostly equigranular or inequigranular. Although K-feldspar megacrysts are not particularly common they occur in some rocks and may be pink, white or grey in colour. Many plutons are texturally heterogeneous, having a range of equilibrium and disequilibrium textures. The sequence allotriomorphic–two-phase–microgranite is commonly developed. All these rocks may be subject to late stage replacement resulting in albitisation, the formation of alteration haloes round mafic minerals, and various textural modifications resulting in the wholesale replacement of the original rock along zones of microfracturing.

Mafic enclaves

These are present in similar abundance to those in Caledonide I–types But they can be very abundant and often show good evidence of synchronous mafic and felsic magmatism.

Mafic dykes

These are quite common, especially composite dykes which may predate the ring structures. Cone sheets may also be present.

Patterns of occurrence

These granites provide good mapping units, are often associated with major faults and may occur either as isolated plutons, or as chains of nested plutons along a lineament. In some anorogenic settings they can be associated with undersaturated syenites and carbonatites.

Mineralization

They have a distinctive pattern of mineralization with rare earths, tin, tungsten and other metals in association with fluorine-rich minerals.

Geochemistry

They are metaluminous to mildly peraluminous and alkaline to peralkaline in their major element geochemistry. They are also characterised by high levels of zirconium, fluorine and rare earths.

Isotopes

The $^{87}Sr/^{86}Sr$ ratios are quite variable, and because of associated metasomatism, range from very low to very high values.

Sources

Most A–types are probably mantle derived, but some seem to be of mixed mantle and crustal origin.

4.4
The S–Type Family

These granites were first defined by Chappell & White (1974) from the Lachlan Fold Belt of Southeast Australia, where they were interpreted as resulting from the melting of a metasedimentary protolith. This was not a new concept and had been described in many articles on migmatites, and had also provided much of the evidence for the granite controversy of the early and middle part of this century. However, Chappell and White established specific criteria by which granites of this nature could be distinguished from other granites. In the realm of field geology they noted the restricted compositional range from granodiorite to granite, the absence of hornblende and pyroxene as mafic minerals, the presence of cordierite, garnet, sillimanite and muscovite and the absence of any mafic association, whether in the form of plutons, dykes or enclaves. Many, though not all of these criteria have stood the test of time.

It is important to realise that the definition applied only to the granites of the Lachlan Fold Belt in Southeast Australia. In other parts of the world, as for example

Fig. 37. Deformed anatexitic S–type granite with folded relic of metasedimentary material. St Cast Plage, Brittany. Hammer 30 cm

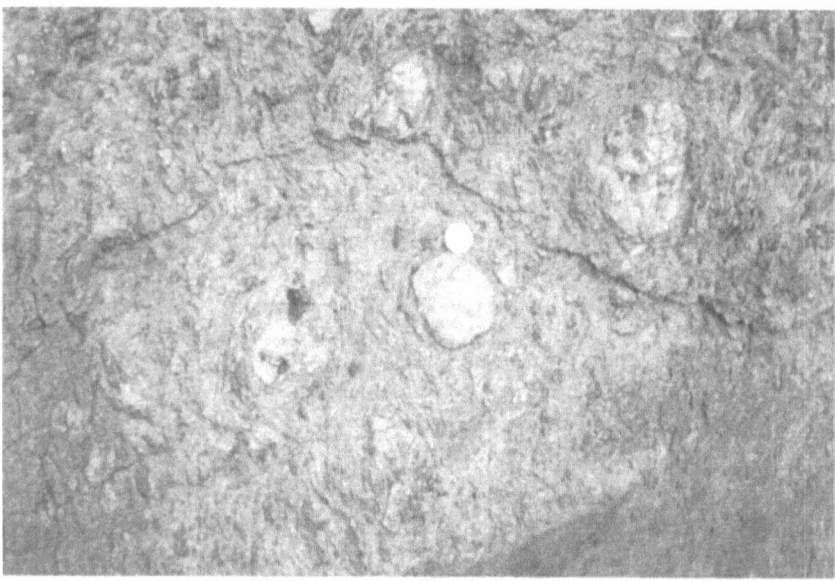

Fig. 38. Foliated anatexitic S–type granite with quartz lumps derived from metasedimentary source rock. St Cast Plage, Brittany. Coin 2 cm

Fig. 39. Relatively undeformed S–type granite with enclaves of predominantly pelitic source material. Beach 2 km west of St Cast, Brittany. Hammer 30 cm

in northern Brittany (Brown, 1979) and in Ecuador (Litherland et al., 1994, Aspden et al., 1995) granites occur which are full of metasedimentary enclaves and which are clearly of anatectic origin, and are, accordingly of S–type in the sense of Chappell and White (1974). Nevertheless some of their geochemical criteria do not conform to those established by Chappell and White. The granitoids from these two areas are extraordinarily similar and differ from the Australian S–types in the same way. The peraluminous index from both regions is above 1.1 and conforms to the Australian criteria, but the Na_2O content is higher and K_2O lower, and these rocks plot consistently in the I–type field for these oxides as defined by Chappell & White (1974). In their paper Chappell and White attributed the low Na_2O values to chemical weathering processes resulting in the removal of sodium into sea water and calcium into carbonates with relative enrichment of the sedimentary pile in aluminium. In contrast the rocks from both Ecuador and Brittany have been interpreted as resulting from the anatexis of immature greywackes (Brown, 1979, Aspden et al., 1995). It is interesting to note that whereas Aspden et al., used the S–type terminology, Brown preferred to call his rocks diatexites of anatectic origin.

Similarly the granites of the Southeast Asian Tin Belt differ from those of the Lachlan Fold Belt, but in the opposite sense to those of Brittany and Ecuador. They are only mildly peraluminous and lie below the 1.1 limit of the peraluminous index established by Chappell & White (1974) as the lowest cut off point for Australian S–types. Conversely their K_2O/Na_2O ratios are identical to those from the Lachlan Fold Belt (Cobbing et al., 1992). The Southeast Asian granites are homogeneous plutons without visible restite. They range from 66% to 76% SiO_2 and the peraluminous index is not higher in the more mafic granites, as is the case with the Australian S–types, but increases with silica content.

Consequently it would seem that S–types, by their very nature are somewhat more variable than I–types. This should not surprise us unduly, as it is evident that crustal sedimentary sources are inherently more variable than the mantle or basaltic lower crust.

The anatectic nature and the metasedimentery character of the source region for the granites from Australia, Brittany and Ecuador is quite apparent, but there are many other granites of a more homogeneous character which, while quite obviously are not of I–type, do not correspond to the definition for the Lachlan Fold Belt S–types. These tend to be biotite leucogranites of various categories and are lacking in other mafic minerals. While they are generally peraluminous they are often not so strongly peraluminous as the Australian S–types although their Na_2O/K_2O ratios are similar. If research on granites of this sort shows that they are neither of I–type or A–type, have crustal isotopic signatures, and that they show some degree of correspondence to the criteria advocated by Chappell and White, it is better to consider them as S–types in the broad sense, but in all probability having other criteria which define them more specifically in a local sense. it is with these reservations in mind that the following guide lines are given. It is quite probable that in the course of time, the nature and variety of S–types will come to match, or surpass that of the I–types.

4.4.1
S–Types

Occurrence and tectonic setting
Collisional settings are said to be most typical for this group, although these are often quite difficult to establish.They occur in the post-collisional and post-orogenic context and also, though more rarely, in transpressional, continental margin, volcanic arc settings. They commonly occur together with crustal I–types in post-collisional/uplift tectonic settings.

Composition
They have a restricted compositional range from granodiorite to syenogranite with monzogranite predominating.

Characteristics of plutons and batholiths
Plutons have a great size range but may be very large. They are almost always simple, being comprised of one characteristic granite unit with a coarse allotriomorphic texture. However the more highly evolved plutons can be very heterogeneous in texture resulting from late stage disequilibrium processes. Batholiths range in size from relatively small linear batholiths to enormous batholiths like the Main Range Batholith of Peninsular Malaysia.

Mafic Minerals
Biotite is usually the only mafic mineral, though muscovite is an important mineral in leucocratic differentiates. In some belts cordierite, sillimanite and garnet may be present in addition to biotite. Actinolitic hornblende may be present in mafic granodiorites.

Textures
Granite plutons are typically characterised by a coarse, allotriomorphic texture. However, in leucocratic varieties there is a range of disequilibrium textures which can be very abundant and results in a good deal of internal heterogeneity. Microgranites are abundant in some plutons and are associated with mineralisation.

Enclaves
Where present these are of metasedimentary rocks, in many cases large quartz nodules are present. Leucogranites are generally devoid of enclaves.

Mineralisation
These granites often have a distinctive pattern of mineralisation with tin, tungsten and rare earths, which occur as disseminations in massive greisens, or in greisen bordered vein swarms. The mineralisation is often associated with extensive kaolinisation.

Geochemistry
These granites are typically peraluminous and high K calc-alkaline, however, tin granites from some belts are not so strongly peraluminous as the Australian S–types. Because of the restricted compositional range lineage is difficult to establish, but differentiation trends within plutons may be either soda or potash dominated. Mineralised plutons are characterised by high levels of fluorine or boron.

Magnetic Susceptiblity
Always low; ilmenite series.

Isotopes
The $^{87/86}$Sr ratios are always high and sometimes very high, generally indicating a crustal origin for these rocks. Nd and Pb isotopes generally support this interpretation when they are available.

Source
Crustal, typically metasedimentary crust, though some authors believe a volcanic protolith may sometimes be involved.

Summary
Although the outline of the geological patterns occurring within the typological system provides a practical framework for the study of granites, many anomalies both apparent and real are bound to occur. Muscovite and garnet, for example, commonly occur in highly differentiated I–type granites, an apparent anomaly which is usually quite readily resolved within the context of their overall granite geology. However, they may also occur in some undifferentiated I–types, and in such cases it is necessary to take into account the whole spectrum of the geological evidence, including the geochemistry and isotope geology. These difficulties emphasize the necessity for the careful description and recording of all aspects of the field geology. However, many of the distinguishing criteria are readily observable in the field. The presence of hornblende and pyroxene in addition to biotite; of mafic enclaves and of mafic dykes, leads to the identification of I–type granitoids. Similarly the identification of a restricted compositional range with no associated mafic intrusives, dykes or enclaves suggests the possibility of an S–type affinity. However, not all granites are so readily identified. Some K-rich I–type and A–type populations are composed predominantly of monzogranite, with only biotite as the mafic mineral. If in addition they lack mafic enclaves and mafic dykes, and have low magnetic susceptibility, it may be quite difficult to decide whether a particular granite is of I–, A– or S–type. Nevertheless, although some plutons are always likely to present intractable difficulties, it is generally the case that most plutons in any belt can eventually be identified as being of a particular typology. Those which elude the net cast by this approach merit particular attention, since they may well owe their individuality to unappreciated geological factors.

5 Field Mapping

5.1
Reconnaissance Mapping: Preparation for Field Work

The essential objective of reconnaissance mapping is that as large an area as possible be mapped to the highest possible standard in the shortest possible time, and in order to achieve this careful planning is necessary. We are here concerned principally with granites, and our first efforts should be directed towards locating all the granite bodies within the area to be mapped, and establishing their degree of accessibility. If a granite is extremely inaccessible it may have to be left for another season when additional time and money is available. In any event it is always best to do the most accessible granites first, because in this way the largest body of data will be established in the shortest time and the learning process will be accelerated.

The procedure is to compile all the available information from previous maps, air photographs and satellite imagery on to maps at an appropriate scale, 1:250 000 is a useful scale for this. All the granite plutons and batholiths and all the rivers, roads and footpaths should be shown. It is also useful to reduce this map to a scale of 1:1 000 000 since this provides a synoptic view of the region, the major geological features and structures, and the distribution of the granites relative to these.

The next step is to plan the operation. This will be based on the map compilation which should show all the batholiths, and at least some of the plutons and their degree of accessibility. A principle to be borne in mind is that the number of geologists to be involved should be kept to minimum. The size of each field party will to some extent be determined by circumstances. If the only access is by foot along small streams, progress will be slow and the field party large. Accordingly it is desirable that the most accessible granites which can be reached from the road be done first. In this case the field party should consist of one senior geologist, one junior geologist and a driver, and they should endeavour to be car-based as much as possible. If they can contrive to keep the car on granite outcrop they will make the best use of their time.

Two such field parties should be able to accomplish the desired objective within the time available. The disadvantage of a larger number is that with more people the information gathered tends to become more fragmented in the course of time, With two or three field parties it is just possible to keep account of the diversity of granite geology. With a greater number it becomes much more difficult.

J. Cobbing: LNES 96, pp. 83–93, 2000.

It is obvious that each field party will have no idea of the granites that the other field party has seen, and in order to establish a unified view it is essential that they get together and compare their work. It is desirable for them to do this at some stage during the field work, but if that is not possible it will have to be done at the end, when all the specimens and maps are back in the office. At this stage all the specimens should be laid out to view in order of plutons, so that all the field parties will be able to compare their own granites with those of the others, and evaluate their degree of similarity and difference. This is a very important process for the integration of the project. It is also at this stage that final selection for geochemical and isotopic analysis is made, which in turn depends on having an overall concept of the work done.

5.2
Use of the Description Sheet

There are two main reasons for using the description sheet. The first is to obtain field descriptions of granites from different field parties which are fully comprehensive and good enough to be compared with one another. The different boxes force the geologist to look for specific features which might otherwise be forgotten. The second reason is that by using the sheet systematically, the observations that the geologist makes improve very rapidly. The geological eye becomes trained to notice features which were formerly ignored, and as a result the quality of observation is much improved. People often believe that the sheet can be completed by a non-geological field assistant. This is a grave error. Using the sheet is a skilled business. No doubt some of the data could be computerised, but an essential feature of its use is the improvement of field observation, and hence improvement in the quality of the data.

In practice the best way of completing the sheet is to describe the easiest minerals first and the most difficult or inconspicuous ones last. These vary from granite to granite, so there can be no set order to follow. The proportion of each mineral should be estimated and entered. At first these estimates will be quite inaccurate, but they will quickly improve and will provide a good basis for the lithological description of the granite. The proportion of quartz, plagioclase and K–feldspar can even be entered on a Streckeisen diagram in the field, which can be extremely useful. When describing each mineral it is best to measure the smallest and largest representatives of each group and estimate the average size visually. The following procedure is recommended for obtaining useful descriptions from primary texture granites, i.e those having a hypidiomorphic or allotriomorphic texture, and for distinguishing them from two-phase granites.

When describing each mineral it is best to measure the smallest and largest representative of each group and estimate the average size visually. In general the most visible minerals are the mafics and these should be described first. It should be established whether one, two or more mafic mineral species are present and

Sample No						Locality							
Rock Type							Granite Unit						

Distinguishing features

Texture and grain size	Primary				Modified		Cataclastic		Incipient Porphyritic	2-phase Equigranular	Microgranite Inequigranular
	V.coarse		Coarse	Medium	Fine						

MAFIC MINERALS	%	Size (mm)		Mode of occurrence			Aligned	Outline	Shape	Colour/Relationships
		Range	Av.	singles	clusters	2-min clots				
Hornblende									Needles Prisms Stubby Equant	
Biotite									Barrels Books Flakes Sheets	
Muscovite									Books Flakes Sheets	

FELSIC MINERALS MEGACRYSTS	%	Size (mm)		Colour	Aligned	Outline	Incl.	Shape/Relationships
		Range	Av.					
K-feldspar								
Plagioclase								
Quartz								

GROUNDMASS								
K-feldspar								
Plagioclase								
Quartz -singles								
Quartz -clusters								

Accessories	Tourmaline Sphene				
Foliation/alignment	Yes No	Weak Moderate Strong	Dip		Strike
Magnetic Susceptibility		Ratemeter Count			

Xenoliths Enclaves	%	Size range		Mafic Yes No	Cognate Accidental	Megacrysts Yes No
	Lithology					
	Shape:	Angular Round Oval	Lenticular	Flattened		
Dykes & veins	Lithology		Width	Dip	Strike	

REMARKS:

Fig. 40. Field description sheet for granites

whether they occur as single crystals, aggregates or both. Next come the megacrysts. If only K-feldspar megacrysts are present, the granite will have an equilibrium texture with anhedral, interlocking grain boundaries and is probably quite a coarse rock, which can be informally considered as a primary texture granite. If the megacrysts are of K-feldspar, quartz and plagioclase, the rock will be a two-phase granite or microgranite and will have a disequilibrium texture, with a finer grained quartzo-feldspathic groundmass.

If the rock is a primary texture granite the next step is to describe the ground-mass, which will consist of quartz, K-feldspar and plagioclase. If the K-feldspar is pink it is readily identified, but if it is white or grey it may be difficult or impossible to distinguish from plagioclase. In this case the two should be grouped together as combined feldspar. K-feldspar megacrysts will probably contain inclusions of mafic minerals and plagioclase, possibly in a zonal pattern and these should be noted. Quartz can be very difficult to describe even though it is often quite visible. It can be present in a variety of colours including blue, grey or brown. The reason for the difficulty of description is that single crystals are the exception rather than the rule. It tends to occur as anhedral aggregates or clusters of several grains, and may often include other minerals such as biotite and plagioclase. The shape and dimensions of these clusters can be extremely variable. In some very coarse granites, the quartz tends to form a discontinuous network of poorly aligned clusters, which may impart a crude fabric to the rock, while in others the quartz may form smaller, inconspicuous, globular or vermiform clusters. Linking quartz clusters are particularly associated with some rapakivi granites but are not confined to them.

When the estimation of the proportions of the major rock forming minerals has been completed they should be totalled up. This total will probably exceed 100% so the original estimate for each mineral will have to be adjusted. The learning curve for the correct estimation tends to be quite steep.

The remaining items on the form should then be completed and the rock classified lithologically and texturally. At the bottom of the form is a box labelled Distinguishing Features in which the geologist should enumerate any features which seem to be distinctive. This makes the geologist observe the rocks in a comparative way and helps in their later comparison.

If the rock contains megacrysts of K-feldspar, quartz and plagioclase it will probably be a two-phase granite or microgranite with a disequilibrium texture, and a different procedure is required. The mafic mineral will almost certainly be biotite and like the felsic crystals it too may well have a bimodal distribution of larger and smaller crystals. All the megacrystic phases should be carefully described and compared with the mineral phases of associated primary texture granite. The ground-mass crystals should then be described and careful note made of the colour , shape, grain boundary relationships of the component minerals, and differences between these and the megacrysts.. The overall texture of the groundmass should then be noted. Having described the rock it is useful to return to the megacrysts in order to observe whether some of the megacrystic phases may be composite. K-feldspar megacrysts may well contain inclusions of biotite and plagioclase, perhaps in zonal form, and resemble the megacrysts of the associated primary texture granite. Most commonly the megacrysts of all species will be dispersed as single crystals, but if they are so abundant that the rock might be called a crowded porphyry, it may be that crystals first perceived as single megacrysts are actually aggregates of more than one mineral. In such cases these will most probably consist of quartz and K-feldspar with a single common grain boundary. This is evidence for the existence of an earlier touching fabric. With care such features can be distinguished during field work but they are normally first seen in thin section.

5.3
Mapping by Textures

In areas of poor outcrop where plutons are not distinguished on aerial photographs or satellite imagery the only practicable way of distinguishing plutons is by their textures. Fortunately the inherent diversity of granites is such that many individual plutons are texturally quite distinct, and these textural differences can be used in the field to distinguish one pluton from another. However, those textures which actually distinguish plutons are equilibrium textures with coherent, interlocking grain boundaries, and are either hypidiomorphic or allotriomorphic. If these textures have been modified or obliterated by later geological processes such as quenching, metsomatism or deformation, their utility is diminished. Consequently the first thing to establish on entering a pluton is whether the textures are primary or not. If they are it is possible to proceed on the lines outlined above. If they are not, the rocks have to be described carefully, and the traverse continued in the hope that the eventual relationship to a primary texture precursor will eventually be revealed.

In the case of granites which have been modified by deformation it is quite common to find augen of granite relics in which the original primary textures are preserved. By observing these a good idea of the texture of the original rock can be gained, and by noting local differences in the intensity of deformation, the textural picture can be amplified, until finally an example of the undeformed original rock may be found. It is quite common to be able to trace a sequence from ultramylonite with rounded relics of quartz and feldspar, through augen gneiss to foliated or unfoliated granite.

With magmatic variants the rocks present a rather different aspect. Two-phase variants and microgranites are quite commonly developed in marginal zones and they may well be the first granites to be encountered. Microgranites usually have a sparse population of megacrysts of quartz, K-feldspar, plagioclase and biotite set in a fine quartzo-feldspathic matrix, and they generally have sharp contacts against two–phase variants or, more rarely, primary texture granites. Two-phase variants have a similar mixture of megacrysts and groundmass, but the proportion of megacrysts to groundmass is very variable and very occasionally they may contain larger inclusions of primary texture granite which can eventually be related to the main primary texture unit. The actual field relationships are often very complex and in the field will appear to be gradational. Nevertheless, the association of primary texture granite–two-phase variant–microgranite is common in certain granites, and provides the geological basis for addressing the field geology of these rocks.

The actual field procedure for describing deformed granites or magmatic variants is the same as for primary texture granites, namely, locate the outcrop on the map, describe the rock using the description sheet or notebook, classify as appropriate and note any similarity or relationship to other granites.

5.4
Field Procedures

1 First of all examine the local outcrop, and any boulders which may be present to see whether the textures are primary or not and whether the boulders are of the same granite as the outcrop.
2 Scan the outcrop and the boulders in a general way, looking for such features as mafic enclaves, mafic dykes, foliation, layering and any other megascopic features. Biotite is virtually always present. It is helpful to notice if the mafic minerals are in single crystals and whether they are euhedral or poorly defined, whether they are in monomineralic clusters or in combination with other mafic minerals. Such features tend to be specific for granite units or suites.
3 Look at the mafic minerals carefully to see if hornblende or pyroxene are present in addition to biotite.
4 Muscovite is readily identified if present, Some granite suites are characterised by fairly abundant euhedral muscovite crystals, but more commonly it is restricted to pegmatites and aplites and may be developed as an alteration product of feldspars.
5 Observe the colour of the K-feldspar. I–types usually have pink or red K-feldspar but it is often also white or grey. In S–types it is virtually always white or grey. Plagioclase is normally grey or white but in some monzodiorites it may be dark grey or black in colour.
6 Measure the magnetic susceptibility with a kappameter. This will determine whether it is a magnetite series or ilmenite series granite. If there is no Kappameter a small magnet can be used on crushed granite powder.
7 Complete the description sheet and classify the rock on the basis of its mineralogy and texture.

The most difficult terrains in which to map granites are equatorial or tropical rain forest, although any kind of forested terrain may be equally difficult. However other kinds of terrains, notably deserts, although easier to work in, nevertheless present their own particular difficulties. These two situations present the extremes under which most geologists are likely to work, and consequently the procedures advocated here, can in most cases, be adapted for use in other situations.

5.5
Mapping in Jungles

The mapping of a large granite batholith in jungle terrain can be quite a daunting experience. The geologist will certainly be aware that the batholith is large and complex, but probably the only visible features will be water, trees, some granite boulders and perhaps a few granite outcrops. Nevertheless, by carefully describing the rocks, and by a process of the systematic elimination of alternatives, it is possible to make a number of useful provisional statements about the components of the batholith. The following procedure allows for most eventualities.

Isolated plutons are relatively easy to deal with so long as there is reasonable access to them. This is because they commonly form isolated hill masses of restricted dimension. Ideally the pluton should be traversed at its widest point. If it can be done from the road this would be the quickest way, but if not another way must be found. Whatever the mode of entry the field procedure is always the same.

On entering the pluton it is necessary to establish the contact and mark it on a map or equivalent document, and if possible indicate which is the younger of the rock units seen. Once inside the pluton the local outcrop and all the boulders should be scanned to see whether they are all from the same rock unit or not. If they are, a representative outcrop can be selected, and a detailed description of the rock made by completing the description sheet. This will provide a frame of reference to which all subsequent outcrops may be referred.

If the pluton is simple, with hypidiomorphic or allotriomorphic textures, all subsequent outcrops will be similar to the first, varying only in such features as the abundance of mafic minerals and K-feldspar megacrysts if present. The size of these minerals does not usually vary very much, though it can do in some cases. The traverse is completed by making careful observations of representatives or variants and by completing additional description sheets as necessary. Representative samples should be collected for analysis if possible.

If the pluton is composite the traverse will cross several rock units. At each contact the field relationships need to be established and marked on the map. As each new rock unit is traversed the procedure for the first is repeated until the traverse is completed. If the pluton is long and linear it may be possible to make several traverses, identifying the same or additional rock units, and thus building up a picture of the pluton as a whole.

Very often in jungle terrain it is necessary to make traverses up streams, because these provide the only feasible access. These generally have some outcrop, but if this is scarce it is usually possible to work with boulders. These will have been derived by tropical weathering along joints and will have come downhill to occupy the valley floor. If all the boulders in the stream are the same it means that the entire drainage basin is underlain by one rock unit. Consequently, by observing the boulders and making careful descriptions using the description sheet, the nature of that rock unit will be established.

If however, there are two or more sets of boulders in the stream, the drainage basin is occupied by more than one rock unit. In this case it is necessary to walk up the stream until only one set of boulders is present, and mark in an approximate contact on the map.

In many cases it may be impossible to make a complete traverse across a pluton. However, sometimes a granite will make a hill mass which is surrounded by a road, and it may be possible to make partial traverses from different directions. These should make it possible to evaluate the geology of all the drainage basins, and construct a map of the pluton.

If a large composite batholith has to be tackled the mapping problem will be more difficult, because although the outer contact of the batholith against the country rock may be well controlled on air photographs or imagery, the inner contacts will be poorly defined or unknown. It may be possible to infer the dimensions of some of these by the curvature of the outer contact, or some other irregularities in shape, but the only satisfactory way of identifying the plutons is by field work. This will entail entering the batholith by as many roads, footpaths and stream courses as possible.

In general it is wise to avoid large rivers since these will drain the entire batholith, and boulders and pebbles of all the rock units will be mixed together in hopeless confusion. It is better to look for small mountain streams which drain a single drainage basin of limited extent. By doing this systematically, and following the same procedure as that recommended for isolated plutons, it is possible to build up a picture of all the plutons which comprise the batholith, together with all the variants.

In these modern time satellite navigation systems are widely available. For those who have them most of the following will be redundant. For those less fortunate, or even perhaps for those whose system has turned out to be non functional, the following comments may be useful. In jungle terrains satellite navigation systems are often affected by the forest canopy which impedes or blocks out the signal. It is a wise precaution not to become too dependent on this technological miracle.

Navigation in jungles is always very difficult. Even with a good quality topographical map it is often difficult to locate ones position, and without one it is much worse. If working from a car along a road shown on a map or air photograph, finding the position is relatively easy. If however, the road is unmarked, it is very different indeed. Satellite navigation systems are a boon in this situation. Otherwise, the only practical procedure, is to record in the notebook the localities seen, by the number of kilometres measured between known points of entry and exit to the traverse, and between each outcrop, combined with changes of direction recorded by the compass.

If on a foot traverse and having a topographical map or air photograph, positions are identified by locating confluences with tributary streams. If neither map or photograph are available it will be necessary to march on a compass bearing and count the number of paces between each change of direction. The traverse will then have to be plotted up that evening on graph paper. It is better to be able to delegate this task to a reliable field assistant if at all possible.

In jungle terrains it is normal practice to record each outcrop seen by a locality number in the field notebook, and give a description of the lithology and geological features in conjunction with the completion of the description sheet. It is possible to proceed in this manner because outcrops are generally scarce, and there is a natural desire to extract the maximum possible amount of information from each one. If the outcrops are complex field sketches should be made and photographs taken. Photographs of the main granite types should always be obtained if possible. At the end of each days field work it is essential to summarise the geology seen, in order to integrate it with the previous days work, and to provide a basis from which to proceed upon the following day. This is best done by compiling a sketch traverse showing the lithology and texture of each outcrop by a chosen colour, perhaps in combination with an appropriate ornament, and a diagramatic representation of any geological particulars, such as contacts, veins, deformation etc. The magnetic susceptibility can be indicated and the provisional rock unit to which it has been assigned can also be shown. Having done this it is useful to write a summary of the days work in which the possible relationships of the units seen are given, and any problems arising from this or former work are enumerated. This procedure is an essential practice for the collation of the field data in a useful and practical way.

5.6
Mapping in Deserts

One of the chief differences between working in deserts rather than jungles is that not only is access generally easier because unimpeded, but it is also possible to do a great deal more useful preparatory work on air photographs and satellite imagery. By using these materials it is possible to identify all the isolated plutons and also many of those within composite batholiths. This is particularly so if these are of Andean I–type with an extended compositional range, because the mafic components are commonly readily distinguished from the felsic on air photographs. These will also sometimes distinguish extremely subtle differences. The result of this is that it is possible to begin field work with quite a reasonable geological map, on which the position and outline of many of the plutons will already be shown, and many of these lines will survive on to the final published map. This does however tend to induce a false sense of security and it is best to embark upon field work with a completely open mind. Nevertheless the planning of field work and the necessary traverses is relatively straightforward.

Because the outline of the plutons is in many cases already known, it is not apparently necessary to distinguish them by the careful examination of their textures as in jungles, and one result of this is that the rocks in deserts are often not actually examined as carefully as those in jungles, where the only way of distinguishing different granites is by their differences in lithology. Paradoxically this can result in a situation where the map itself is very good, but knowledge of the granite units is relatively poor. For this reason it is necessary to make a particular effort to obtain detailed descriptions of the granites by the systematic use of the description sheet and the notebook. This is not easy because with virtually continuous outcrop it is impossible to record every traverse comprehensively. Nevertheless, by adopting a systematic procedure akin to that used in jungle terrains it is possible to develop a satisfactory compromise which allows the work to go forward at a suitable rate.

On entering any new pluton or rock unit it is best to look around in a general way to establish whether the granitoids are of a primary texture or not, and then, having selected an outcrop which seems to be representative, make a detailed description of it using the description sheet. This serves to define the character of that particular rock unit until something different is seen. Having done this it is now possible to proceed more quickly, scanning the rocks continually to see if there is any departure from the model established at the first outcrop. Any such differences should be carefully described in the same way until the traverse of that pluton is completed. In this way it is possible to cover long distances while at the same time ensuring that proper descriptions of the lithology and granite geology are recorded.

The notebook can be used in a similar way to the procedure advocated for jungles. The compilation of a sketch traverse of each days work, followed by a summary discussing the affinities of the units seen, and any problems identified, provides a geological framework on which to begin the following days traverse.

Navigation in mountainous deserts does not present any problem since there is abundant topographic detail on maps, air photographs or imagery which render position finding a relatively easy matter. In arid peneplaines however, there are few features which enable precise, or even approximate location of position. Although such features may be visible on the air photographs, it is often very difficult to locate them on the ground. Satellite navigation systems are invaluable in this situation, but if these are not available, it may be necessary to resort to pace and compass methods of field work.

5.7
Collection of Samples

The underlying constraint for all regional mapping in areas outside Europe, North America and but a few other countries, is that any outcrop will only be visited once. The controlling factors of inaccessibility, logistics and funding ensure that there will never be another opportunity to return and check observations, or to collect better specimens. This means that hand specimens for reference, and larger specimens for geochemical and isotopic analysis have to be collected during the course of field work, and the time and effort required for this should not be underestimated.

Reference hand specimens should be as fresh and unweathered as possible, but it is better to collect a weathered sample rather than none at all, because at least the texture and mineralogy should be visible. In order to collect decent specimens of granite a sledgehammer of 8 or 10 lbs weight should be used. An ordinary geological hammer is quite inadequate.

It is also necessary to collect larger samples for geochemical and isotopic analysis. Between six and ten samples should be taken from each pluton if possible. Sometimes it will not be possible to achieve this because of the poor condition of the rock through weathering or alteration, but in most plutons it is possible to get something, and even one analysis from a pluton provides information which otherwise would not be available.

Samples should be completely fresh and unweathered and at least 5 kg in weight. This should be enough to provide chemical and isotopic analyses with enough left for a hand specimen and a thin section.

6 Handling the Data

6.1
Selecting Samples

The first, most necessary and most important part of the data handling process takes place at the end of each field season. This has two objectives, the first is to thoroughly familiarise each field party with the granites that the other field parties have mapped, and the second and most important, is to make a detailed comparison of all the granite plutons, their component units and variants, and decide on the basis of the field evidence and the assembled collection, the relationships of the various plutons to one another. This is the first comparative correlation, others will follow.

The next step is to select specimens for geochemical and isotopic analysis. The number of samples of geochemical quality is bound to be uneven. Some plutons will have a good many, others hardly any or none. However, it is a good general principle to obtain analyses from as many plutons as possible, since this spreads the basis for comparison. If, for any reason, a pluton which seems to be important is under represented, it is better to obtain an analysis from the best material available rather than have none. Similarly if a pluton is so highly deformed that it is an orthogneiss, it should be analysed chemically and isotopically in the normal way. Results from such highly deformed plutons are often extremely useful.

Once the samples have been selected they have to be crushed. The usual procedure is to put them first through a jaw crusher and then through some kind of mill. The jaw crusher accounts for a certain amount of contamination from iron, which is negligible. The next step in the ball mill is important because this is when contamination can occur. It is best to use an agate mill if possible since, although there will be contamination from silica, there will be none from other elements. The samples should not at any stage be put through a tungsten carbide temer since this results in contamination by tungsten and other metals, which are precisely those which need to be determined accurately.

The samples can then be analysed by XRF or ICP methods for major elements and the required selection of minor and trace elements. Splits from the same powders can be used for whole rock Rb/Sr dating and also Sm/Nd studies.

While the analytical process is going forward, thin sections can be made from a piece of the original sample. These should be large, because granites are coarse grained rocks and a standard thin section might only have a few crystals, yielding

J. Cobbing: LNES 96, pp. 95–101, 2000.
© Springer-Verlag Berlin Heidelberg 2000

very little information. The sections can be described systematically but as briefly as possible and with special regard to their textural features, including magmatic and dynamic overprinting, and with careful notation of the accessory minerals. Polished sections of selected samples enable the opaque minerals to be determined.

Once the sample preparation and thin sections are complete, the entire collection, together with thin sections and maps, should be gathered together in one place so that there can be continuous interaction and reinterpretation between the different data sets.

6.2
Essential Geochemistry

With the range of trace elements currently available with modern analytical methods it is possible to embark upon a very sophisticated programme of geochemical interpretation. However, in a regional project there will be a large number of analyses from numerous plutons, so that a programme has to be devised which is both practical and useful. In practice there are three major objectives to be addressed, classification, comparison and differentiation, and three lesser objectives, typology, source region and tectonic setting.

6.2.1
Classification

The lithological classification of Streckeisen (1976) is achieved by plotting the normative values on a QAP diagram. This is a simple and practical diagram but it suffers from the disadvantage of all ternary diagrams that with large numbers of analyses the points become become too closely crowded to be readily distinguished. This problem can be overcome by using the Q' ANOR diagram of Streckeisen & LeMaitre (1979) or the nomenclature diagram of La Roche (1978) and Debon and Le Fort (1982). However, the Streckeisen triangle remains universally popular.

The aluminous index of Shand (1947) is calculated directly from the major element data and can either be directly plotted in binary diagrams as, for example against SiO_2, or it can be incorporated in the characteristic minerals diagram of Debon and Le Fort (1983) in which it is plotted against a measure of the mafic minerals.

The alkali balance can be determined for I–type suites with an extended compositional range by using the calc-alkali index of Peacock (1931) in which CaO and combined alkalies are plotted together against SiO_2. the two lines should intersect at between 55% and 61% for calc-alkali rocks. Those which intersect at more than 61% are termed calcic, those between 55 & 50% are alkali-calcic and those below 50% are alkalic. For suites dominated by monzogranites this method is inappropriate because the lowest SiO_2 values are likely to be greater than 65%. Consequently the diagrams of Irvine and Barager (1970) and Kuno (1969) where $K_2O + Na_2O$ are plotted against SiO_2 to define the fields of Alkaline, Calc-Alkaline and Tholeiitic granites are more useful. Other diagrams such as that of Wright (1969) and the CaO Na_2O K_2O ternary diagram can also be used. Additionally potash-rich suites can be distinguished on the K_2O vs SiO_2 diagram of Peccerillo and Taylor (1976) and Atherton (1979).

By using these diagrams the principle major element characteristics of the rocks will have been established which, in many cases should also reveal the lineage and line of evolution of certain suites.

6.2.2
Comparison

For a regional project this is probably the most important aspect of the geochemical study. The simplest and most effective way of doing this is by the use of single element plots. These are made by plotting the value of each element from the analyses of each pluton on a horizontal line (Cobbing et al., 1992). This enables the abundance of each element to be viewed as a whole for the entire population, and the distribution and variation of every element in all the plutons is immediately visible. A more sophisticated method of comparison is by statistical methods, whereby certain groups of elements which are statistically correlated, are grouped together by cluster analysis into groups of related plutons. If this method is adopted care should be taken to exclude all two-phase variants, microgranites and other highly differentiated rocks since the differentiation patterns will have swamped the background geochemistry. If only primary texture granites are used a useful degree of correlation can result.

Each granite unit, super-unit or super-suite has a distinctive chemical character consistent with its derivation from a particular source region. This means that throughout their compositional range distinctive compositional properties persist, for example high or low Sr, Ba, Cr, Na_2O and CaO etc., which reflect similar features of their source rocks. By scanning single element plots such features can be readily identified, and these can be plotted against SiO_2, Thornton and Tuttle Differentiation Index or any other index which will reveal their consanguinity. In the Lachlan Fold belt Chappell (1984) has illustrated the geochemical distinction of the Bemboka and Glenbog suites of the Bega Batholith by Harker plots of Sr vs SiO_2 and similar plots have been used for distinguishing other suites in that belt (White & Chappell, 1983). Atherton et al. (1979) used K/Rb vs TTDI and Zr and Y vs SiO_2 in order to distinguish the Santa Rosa Super-unit of Nepeña from that of the Rio Huaura in the Coastal Barholith of Peru.This demonstrated that the provisional field-based classification was erroneous. Subsequently Atherton & Sanderson (1985) utilised Sr, Rb, K/Rb, Ba, Th, Pb, P_2O_5, Y, Zr, Zn, Cr, V, Co, La, and Nd vs TTDI to distinguish between the several super units of the Arequipa segment. They also further characterised the Nepeña and Huaura super units of the Lima segment by plotting K_2O, Na_2O, MgO, CaO and total iron vs SiO_2, and Rb, Sr, Ba, Th, Zr, and Y vs TTDI.

6.2.3
Differentiation

This is a feature of most granites, and as it is generally the essential precursor to any mineralisation, it is vital to understand which of the several possible paths of differentiation is represented. The most commonly used diagrams are Harker plots where oxides are plotted against SiO_2. Soda and potash usually increase with increasing abundance of silica, while other oxides such as iron, magnesia and lime decrease. Minor elements and ratios can be plotted in the same way, and also against the Thornton and Tuttle Differentiation Index, or against other indices such as K/Rb or TiO_2. Differentiation trends in granites tend to be dominated by a preferred enrichment in K_2O or Na_2O, as these oxides generally increase with increased SiO_2 values. The trends for most elements are generally regular, but in the most highly differentiated rocks they become extremely irregular, or even chaotic, and this probably reflects the supplanting of magmatic differentiation by hydrothermal activity.

6.3
Typology

In general most I–type granites are metaluminous and most S–types are peraluminous, but there are many exceptions. The most highly evolved representatives of both are peraluminous and can only be distinguished by their lineage. The characteristic minerals plot of Debon and Le Fort (1983) illustrates this point quite clearly. This index should only be applied to undifferentiated granites.

For a given content of SiO_2 S–types will have lower CaO, Na_2O and Sr than I–types, and this is the basis for their distinction by the Na_2O vs K_2O diagram of Chappell and White (1974). They tend to have higher K_2O and Rb for a given SiO_2 content and are generally higher in K, Rb, Cr, Ni, P_2O_5 and Fe_3/Fe_2. In S–types P_2O_5 increases with differentiation, whereas in I–types it decreases. Th and Y increase in I–types and are constant in S–types, La and Ce decrease in S–types and are constant in I–types (Chappell & White, 1992).

A–types are generally similar in their major element chemistry to I–types, but are distinguished from them by their higher values of Zr, Nb, Y, La, Ce, Sc, Zn, and Ga. In fractionated S and I–types Nb and Ga move towards A–type values as does Y for I–types. The most widely quoted discriminant for A–types is the Ga/Al ratio (Whalen et al., 1987).

6.4
Tectonic Setting

Much has been written on the relationship of the composition of igneous rocks to their tectonic settings. With volcanic rocks the geochemical criteria for distinguishing oceanic, island arc and within-plate settings seem to operate with reasonable success (Pearce & Cann, 1973). A similar attempt for granites was made by Pearce et al. (1984) and Harris et al. (1986) in which ocean ridge granites (ORG), volcanic arc (VAG, syn-collisional (Syn Colg) and within plate (WPG) granites were distinguished on the basis of certain trace elements, principally Rb, Nb and Y. There is a certain, though not complete, correspondence between the identified settings and the typological view of granitoids. Thus the M–types correspond to ORG, Cordilleran I–types to VAG, S–types to Syn C0LG and A–types to VAG. There is no slot for post-collisional/uplift I–types which, however, plot in the VAG field. Many orogenic and post-orogenic granitoids plot in an area spanning the triple point of all three fields.

These diagrams are useful though not definitive, and care has to be exercised in using them. As in all studies related to granite origins and source, it is necessary to exclude all two-phase granitoids, microgranites and other highly differentiated rocks since their source signature will have been diminished by fractionation processes. Differentiated I–types and S–types tend to plot in the WPG field. Nevertheless these diagrams can be useful and help to indicate unusual features of certain granitoid suites.

6.5
Isotope Studies

A very wide range of granite-related problems can be investigated by isotopic means. However, for practical reasons a limited approach has to be adopted for reconnaissance work, and this will focus particularly on determination of the age of intrusion and of the age and composition of the source region. It is as well to bear in mind that good material collected for geochemical and isotopic work can in many cases be used for further laboratory studies not envisaged at the time of collection.

In favourable circumstances a minimum and a maximum age for a pluton can be established by the geological relationships. Thus, if a pluton is seen to cut a stratigraphic formation the age of which is known, a minimum age for the pluton is also known. If in addition it is overlain by another formation of known age the age of the pluton may be quite well constrained by the stratigraphic evidence. This situation however tends to be rather unusual, and in most plutons their age is only poorly constrained by the stratigraphic evidence. Isotopic dating by various methods can result in very precisely determined ages for some plutons, though in some cases complicating factors are present which have to be taken into account. The main methods currently in use for the isotopic dating of rocks are the Potassium/Argon method, the Rubidium/Strontium method and the Uranium/Lead method on crystals of zircon or monazite.

6.5.1
The Potassium/Argon Method

Ages are obtained from granites by separating mafic minerals such as biotite and hornblende and measuring the amount of argon produced by the radioactive decay of potassium. In many cases this can give good results. For example the results of the first geochronological study on the Peruvian Batholith by Wilson (1975) correspond very closely to those of later Uranium/Lead studies by Mukasa & Tilton (1985), whereas the later Rb/Sr work did not correspond as well (Beckinsale et al., 1995). Unfortunately however, argon is easily lost from the rock by subsequent heating, and this can result in the resetting of the isotopic clock and the production of much younger ages. In many cases such resetting is not at all apparent, and may only be discovered when another isotopic method, not subject to the same difficulties, is subsequently employed. This was the case in Southeast Asia where granites of known Jurassic age gave Cretaceous ages by the K/Ar method (Bignell & Snelling, 1974, Beckinsale, 1979).

Potassium/Argon dating is done on minerals extracted from a single rock specimen. If both biotite and hornblende are present each can be determined separately, and if these ages are close it is a good indication that the age obtained may be reliable. If there is a discrepancy, biotite usually has the younger age as its crystal structure is not so retentive of argon as that of hornblende. In such cases the figures obtained for hornblende may be more reliable.

Potassium/Argon geochronology is a useful method for reconnaissance work because an age can be obtained from a single sample. It is more useful in I–type granitoids than S–types because of the possibility of obtaining ages on a biotite–hornblende pair.

6.5.2
The Rubidium/Strontium Method

The age of the rock or mineral is obtained by measuring the quantity of Sr^{87} produced by the decay of Rb^{87}. Although this can be done on individual minerals the usual practice is to produce a whole-rock isochron. This involves collecting 5–10 large samples from each pluton, crushing them and analysing the powders. This should result in the production of an isochron in which all the results lie upon a line, the slope of which indicates both the age of the rock and the initial $^{87}Sr/^{86}Sr$ ratio. This latter figure is indicative of the composition of the source region. Ratios of 0.704 or below indicate a mantle source, those up to 0.708 a mixed mantle–crustal source and those of 0.710 and above a wholly crustal source.

The method is widely used in granite studies and it has contributed greatly to modern interpretations of granite geology. However, the Rb/Sr systematics are subject to disturbance and in many cases the results obtained are scattered and do not lie upon a perfect line. This situation is termed an errorchron and although the result is not useful for establishing the age of the granite, the initial ratio obtained may be useful.

6.5.3
The Niodymium/Samarium Method

The isotopes of these elements can be obtained from the same powders as those used for Rb/Sr geochronology and so there is good reason for the two methods to be used together. For granites of Phanerozoic age the results are not useful for obtaining a reliable age for the pluton, but they provide a good idea of the isotopic composition of the source region which, in combination with the data for Sr can be used to calculate the proportion of mantle vs crustal material in the granite. In addition the results may indicate the age of the source region, and although this is a somewhat theoretical concept, it often provides an additional item of useful information for the interpretation of the origin and history of the rock.

6.5.4
The Uranium/Lead Method

In this method zircon or monazite crystals are extracted from the crushed powder of a single rock and are sorted manually into different populations on the basis of their crystal morphologies. The most suitable crystals are selected and dissolved, and the proportions of radioactive lead derived from the decay of uranium is measured in a mass spectrometer. The results are plotted on a concordia–discordia diagram and the intersection of the resulting line with the concordia curve gives the crystallisation age of the crystal, and also the age of the source region from which the crystal was derived. (Liew & McCulloch, 1985). Ages produced by this method can be very precise but it is extremely laborious and expensive. Because zircons are extremely refractory, it is possible for a crystal to pass through more than one magmatic–sedimentary cycle, and the rounded cores of some euhedral crystals reflect this process. These cores can also be dated by the latest single crystal technologies, which can then reveal the nature of the crustal processes in considerable detail. In some cases a zircon crystal may have several well defined zones in addition to a central core. In favourable circumstances each zone can be dated by using laser technology and the history of events affecting the granite source region can be obtained (Miller et al., 1992).

Thus the isotopic methods currently available provide powerful tools for obtaining the ages of granite plutons, and for forming a good idea of the age and composition of their source regions. This information, when used in conjunction with the whole-rock geochemistry, can provide a very complete picture of the history and evolution of any granite body. However, the difficulty and expense of these methods means that at present they are only called into full operation for very specific targets. For reconnaissance mapping a judicious mixture of K/Ar, Rb/Sr and Sm/Nd methods is probably the most suitable, with U/Pb being held in reserve for the most intractable or most important problems. Properly conducted reconnaissance and regional studies succeed in identifying those granite plutons which, in some way or another, can contribute to a resolution of the many problems which are still outstanding in the overall study of granite geology.

7 Integrating the Data

7.1
Storage of Specimens

During the course of field work on a regional project a large number of specimens will have been collected, which will form the basis of all future work and interpretation. For this reason they need to be readily accessible, so that they can be constantly viewed with respect to additional data as it becomes available. In order to achieve this the specimens, thin sections, maps and geochemical and isotopic data all need to be in the same place, namely the working environment of the project geologists.

It is impossible to collect small pieces of granite which show their essential properties. Because of the nature of granites many specimens are large and will not fit into wooden cabinets. For this reason metal racking is the best way of storing specimens. This has the advantage that the the spaces between the shelves are adjustable, and also that because the racking is open, the specimens are on constant display and the samples are easy to find, although they may get a bit dusty. The racking should be placed at one end of the work place, and close by there should be a set of drawers containing all the maps with all the specimen locations marked. Never store the specimens in a dark basement miles away from the work place.

7.2
The Workplace

The ideal working environment for a geological project is to have an open central space in which all the material relating to the project is located and can be worked on. This should include the specimens, thin sections, maps, air photographs, satellite imagery, stereoscopes and microscopes. It should also have a sink with running water, The workplace of each geologist should be located around this space as small offices or cubicles equipped with a desk, bookcases or shelves, telephone, microscope and personal computer, This arrangement should ensure the seclusion necessary for writing, combined with the maximum opportunity for constant reference to the data and interaction with colleagues.

J. Cobbing: LNES 96, pp. 103–108, 2000.
© Springer-Verlag Berlin Heidelberg 2000

7.3
The Portrayal of Granites on Geological Maps

Granites have been portrayed on geological maps in a variety of different ways and most of them are unsatisfactory. Some maps have the granites coloured with the same colour as that of other units of similar age, e.g. green for Cretaceous, blue for Jurassic etc., and are distinguished only by an ornament. As a result it is practically impossible to see the granites on the map, since they do not stand out against the background. Other maps show the granites coloured with respect to their perceived structural or tectonic significance. Since these judgements are mostly subjective this kind of presentation results in a semantic tangle from which it is impossible to escape. The chief principles underlying the presentation of granites on maps are:

1 The divisions chosen should be selected as objectively as possible.
2 The system of presentation should be as simple as possible.
3 The colours chosen should be distinctive and uniform for granites of all ages. Where granites of several different ages occur on the same map sheet it is possible to consider the representation of the older granites in darker shades of the same colour as the younger ones.
4 The map ought to convey as much information about the geology of the granites as possible.

People who use the map want to know certain specific things such as, the lithology and range of lithologies, what mafic minerals are present, whether K-feldspar megacrysts are present, the texture, whether the granite is foliated; whether it is mineralised and if so, what are the metals? These are all objectively determined criteria which are of permanent value, and which allow the derivation of alternative systems from the map for diagramatic representation. The following scheme is recommended to meet these criteria.

The range of lithologies to be portrayed is gabbro, diorite, monzodiorite, tonalite and trondheimite, granodiorite, monzogranite and syenogranite. It is better if these are all shown in shades of pink or red, though gabbros can be shown in some other distinctive colour such as green, blue or purple. Diorites can be shown in the darkest shade of purplish red, and perhaps monzodiorites could be indicated by a sprinkling of an ornament such as a small k. Tonalites and trondheimites should be in a dark shade of pink and granodiorites in a lighter shade. Monzogranites and syenogranites should be a bright red. These four colours should be adequate to portray the range of lithologies present.

Each pluton should be named and the age of the pluton placed in brackets after the name. It is better not to write on the pluton itself as this obscures the geological information, but at the side with an arrow pointing to the relevant pluton. This obscures other geological information but our concern is with the granites. Age can be shown as e.g. 125 Ma if isotopically determined and the isotopic method used e.g. K/Ar also indicated; or by a stratigraphic symbol such as K for Cretaceous if the age of the pluton is stratigraphically constrained, but for which there is no isotopic data.

It can be assumed that if a lithology is shown by a colour without any qualifying ornament, that it is equigranular with a hypidiomorphic or allotriomorphic texture.

It can also be assumed that all monzogranites, syenogranites granodiorites and tonalites contain biotite. If they contain hornblende or pyroxene as well, this can be shown by a small dark prismatic ornament.

If the pluton contains K–feldspar megacrysts these can be shown as small open rectangles or circles.

If parts of the pluton are microgranitic or have two-phase textures this can be shown by a fine stipple. In some cases it may be necessary to mix this stipple with some other ornament.

If the pluton is foliated this should be indicated by fine flowing lines following the trend of foliation. The dip can also be indicated by a tick and a value if known. The sense of vergence can also be indicated if known.

If dykes are present these should be portrayed by a single coloured line, green for mafic dykes, red for acid.

Mineralisation can be indicated by placing the symbol for the appropriate metal to follow the name and age of the pluton.

It is unfortunately not possible to portray finer geological detail at map scale and other important information such as the presence of enclaves, layering etc. will have to be reserved for the report.

7.4
Writing the Report

Every geological report is bound to be different. The following recommendations are made to encourage orderly presentation.

By the time the report has to be written, the geologist will be completely familiar with the data and will be aware of the overall geological and geochemical parameters. It will be known which plutons are normal and which are not, which are mineralised and which are not, and in all probability the reasons for these differences will be understood. The difficulty is presenting the data as comprehensively but as concisely as possible, and the best way to achieve this is to make the maximum use of diagrams and tables. This enables the interested reader to locate the information quickly without wading through descriptive pages of text in which the data lies buried. For the same reason written descriptive passages should be kept as short as possible. All geochemical, isotopic and other numerical data should go in the appendices.

Before starting it is beneficial to write down a list of all the known features of the granites, and especially a list of all the problems identified. This helps in the the organisation of the report into descriptive and interpretative sections. Furthermore the writing of the descriptive passages clarifies the mind, so that by the time the summary and conclusions have to be written, there is a firm base upon which to proceed.

The data should be presented under the headings of the main methods of investigation employed, e.g. geological, geochemical and isotopic. A brief introduction should include the history of former work, and a general outline of the geology of the region and its tectonic setting. The basis for the scheme of presentation should also be outlined, and for ease of reference this scheme should be the same in all the sections.

7.4.1
Geology

The geological presentation of plutons in geographically or geologically defined sub-areas should include a brief description of the field characteristics of each pluton and its geological relationships, followed by a description of the pluton itself, a brief statement of the texture and mineralogy of each unit and its variants. This can be accompanied by a table based on the description sheet or, if there is insufficient room, the textures and mineralogy can be summarised in a general table for all the plutons within a particular sub-area.

7.4.2
Geochemical

The first aim should be to classify the granites by their major element criteria, and similarities or differences indicated on this basis should be discussed here . This should be followed by a section considering minor and trace elements using binary plots or spidergrams as needed. Both major and trace elements should be used to evaluate the questions of typology, lineage, differentiation and actual or potential mineralisation. It is easy to write a lengthy and detailed geochemical section because the data is capable of revealing extremely subtle information about granite relationships. Nevertheless it is better to write a straightforward and reasonably comprehensive account rather than to try and cover every possible eventuality. If the data is given in full in the appendices, other geologists with more time will have that opportunity, and they will also be able to compare the results with those from other granites.

7.4.3
Isotopic

This section has fewer problems of presentation because the results are more precisely focussed, and generally allow for a relatively limited degree of latitude in their interpretation. They do however, have a very precise relationship to some areas of difficulty in the geological and geochemical sections, which can be addressed in the summary.

7.4.4
Summary and Conclusions

Having completed the factual presentation, the original list of preliminary conclusions can be returned to, and modified in the light of the written document. This modified list can form the basis for the summary chapter.

7.5
Archiving the Data and Material

The first step towards ensuring that the proper archiving of granite specimens is done is during during field work. First of all the locality is identified on the map, air photograph or whatever topographic material is available. A locality and/or specimen number is assigned to it and this is written on the map at the sample locality. It is then copied into the field notebook, together with the map reference or other coordinates, which is then followed by a description of the locality and of the sample. When the sample has been taken the sample number is written on it with a marker pen, preferably in two places. The sample is then put in a plastic bag, and the sample and locality numbers are written on the outside of the bag. At the same time the sample number and the grid reference or topographic coordinates should be written on a piece of card or paper and this should be placed inside the bag with the sample. Having taken all these precautions it is still quite common for specimens to go missing or, for some unaccountable reason, not be identifiable.

Considering the time, cost and effort which has gone into collecting granite samples, the preparation of thin sections, the crushing and preparations of powders for analysis and the analytical procedures themselves, it is rather surprising to find that the existence of such a body of material is often not regarded as a scientific and human resource, but simply as so much useless rock taking up too much valuable space. Often enough, valuable collections from remote and inaccessible regions have been quietly disposed of on the whim of some administrator, or even the office cleaner, while the geologist who collected them is away on field work. This would never happen if they were collections of dinosaur bones or other palaeontological material, but granites, for some reason, are often just regarded as material which can best be used as aggregate for concrete. Many granite geologists have suffered from this attitude and many valuable and irreplaceable collections have been casually disposed of. Consequently it is necessary to take specific measures for the archiving and preservation of these collections.

One of the problems is that granite specimens collected for geochemical and isotopic analysis are large and do not fit easily into standard drawers. it is often necessary to reduce the size of large single specimens to a smaller size in which case each piece needs to be properly numbered.

Lists of the samples collected should be compiled as early as possible. This is easily done on field work at the end of the day since only a relatively small number of samples will have been collected. The list should have the name and number of the map sheet for each sample collected, followed by the map reference and a lithological/unit identification. Spaces for thin sections, geochemical and isotopic analysis can be filled in later as necessary. Thin sections and powders of all analysed material should all be clearly labelled and archived, either separately or together with the rocks. It is also important to preserve the original map sheets with their geological and topographic information, together with the original field note books.

It is important to realise that analytical techniques are improving very rapidly and studies can be done on older collections which can greatly improve the geological interpretation and which, at the very least, avoids the necessity for returning to former localities to collect duplicate material.

Perhaps the most striking example of the value of proper archiving is that of the Burgess shale in the province of British Columbia in Canada. Walcott, who was then the Director of the US Geological Survey and was a very influential man in the scientific community of that time, used to take his summer holidays on horseback in the Rocky Mountains as a break from the cares of office. He discovered the outcrop of the Burgess Shale which is of lowest Cambrian age, and which carries a rich and diverse fauna in a remarkable state of preservation, such that entire fossils, and even impressions of their soft parts, were perfectly preserved. Walcott described and published some of these, but the collection remained undisturbed for almost a century before anybody looked at it again. Between 1971 and 1985 Whittington and his colleagues examined the collection, and found that it contained examples of creatures which were unknown, or even unimagined before, and their results were published in numerous papers summarised in Whittington and Conway Morris (1985). Walcott had formerly described some of these as trilobites and other arthropods, which was incorrect, but understandable considering the burden of administration to which he was subject. At the very least he deserves credit for finding and realising the importance of the locality, and for ensuring the preservation and archiving of the collection. Subsequent to the work of Whittington and his colleagues, others have returned to the area and have found other outcrops of the same Formation which carry an equally diverse fauna. Work on the original and subsequent collections has transformed our knowledge of the origins and diversity of metazoan life during the biological explosion at the beginning of the Cambrian. None of this would have been possible had it not been for the careful archiving of the original material.

One cannot pretend that any granite collection will contain information of such extraordinary value. Nevertheless carefully archived collections are capable of providing additional information for areas which may no longer be accessible.

8 The Granite Controversy and its Aftermath

Opinions about the nature and origin of granite have divided geologists for the greater part of the history of the subject as a recognised branch of the natural sciences. The earliest known controversy in the later part of the eighteenth century revolved around the contrasting views of two schools of thought, the Neptunists and the Plutonists. The Neptunists were represented principally by Abraham Werner of Freiberg 1749–1817, and the Plutonists by James Hutton of Edinburgh 1726–97. The Neptunists believed that granites were a chemical precipitate from a universal ocean whereas the plutonists considered them to be due to the consolidation of matter made fluid by heat. Hutton observed veins of granite which had intruded crystalline metamorphic rocks in the Scottish Highlands, and concluded that they could only have originated by the solidification of molten rock material injected from below. This observation contributed to the eventual resolution of the dispute in the favour of the Plutonists.

The Neptunists cited an example of a crystalline rock containing ammonites, at Portrush in Northern Ireland, in favour of their hypothesis. However its crystallinity was subsequently shown to be the result of thermal metamorphism, caused by the emplacement of the nearby Fairhead sill of Tertiary age and the Neptunist position could no longer be sustained. Many influential geologists however, continued to favour the Neptunist model and it was not until the middle of the nineteenth century that it faded from the mainstream of geological thought.

Field observations by the Plutonists established a body of geological evidence which supported the concept that granites were generated at some unknown depth within the Earth's crust, and were emplaced at higher levels as mobile magmas, which then solidified to form crystalline granite. The observation of sharp, cross-cutting contacts, the occasional preservation of a pre-granitic roof and the recognition of contact thermal aureoles, supported the concept and gave rise to the acceptance of molten intrusive granite. Those adhering to this view subsequently became known as the 'Magmatists'.

At the same time that these concepts were becoming established, geologists working in high grade metamorphic terrains, especially in France and Scandinavia, became aware that there was an apparent continuous transition from rocks which were undeniably granitic, to schists and gneisses which were demonstrably of metamorphic origin (Michel Levy, 1877). There often seemed to be a shadowy zone between rocks of metamorphic aspect and granites, in which granites, gneisses and schists were mingled together forming large areas of lithological and textural complexity, which eventually became known as migmatites (Sederholm, 1907).

J. Cobbing: LNES 96, pp. 109–115, 2000.

There were, however divided opinions upon whether the granitic component were veins related to a nearby granite, or whether they were lateral secretions generated during ultrametamorphism. Although it was accepted that both situations existed, the final position of the Transformists as they became known, was that granitic rocks were generated through ultrametamorphic processes from older metasedimentary and meta-igneous rocks.

The Transformists were represented by geologists who had worked on granitic rocks in metamorphic terrains occurring in France, Scandinavia, Scotland and parts of North America. They considered granites to have formed by some process of ultrametamorphism whereby metamorphic rocks, and especially metasedimentary rocks, were transformed into granites. The magmatists considered granites to have resulted by crystallisation from a liquid magma, and many of them believed that granites were formed by processes of fractional crystallisation from a primary basalt magma.

The application of optical microscopic studies to igneous rocks towards the end of the nineteenth century spread rapidly to every region where granites were studied and was used by geologists from every school. However, the most influential exponent of the technique was Rosenbusch (1877, 1896) who studied the thermal aureole of the Barr–Andlau granite, and concluded that permeation by magmatic juices was impossible. Although many geologists resisted this view on the basis of their field experience and their own studies, the technical advances of microscopy, gechemistry and experimental petrology tended to support the Magmatist rather than the Transformist school. A contributing factor may have been the difficulty of reproducing ultrametamorphic conditions by experimental methods.

During the early part of this century geological studies in the Western Cordilleras of North America began to amplify the area of debate. Iddings (1909) and Daly (1912) recognised that the components of the giant linear Cordilleran Batholiths comprised a series, ranging from diorite to granite, in which the more basic rocks were the oldest components and the most acid the youngest. The mapping also established that later plutons intruded earlier plutons, with no visible sign of deformation, and this observation gave rise to the concept of magmatic stoping. Daly calculated the volume of the different magmas in the crust, and in these Cordilleran batholiths the proportions of intermediate and acid rocks was in far greater excess than that for the associated basic rocks. It became apparent that Cordilleran batholiths differed in many respects to most of the granites which were being studied in Europe and elsewhere, and which were principally of monzo-granitic composition. Thus geologists began to appreciate that there were different granitic series, but they did not know how they were produced or why they were different. Geologists began to talk of Daly Batholiths for Cordilleran granites and Suess Batholiths for the granites of continental interiors. The numerous sharp contacts and differing lithologies of the component plutons of Cordilleran Batholiths left little doubt that they were of magmatic origin, whereas the complex field relationships of granites with their host rocks in metamorphic and migmatitic domains indicated an alternative mode of origin.

There is a vast literature on the differences between the Magmatist and Transformist schools of thought, summarised mainly in the favour of the Transformists by Read 1956, in which he presented a collection of influential papers written between 1939 and 1954 and entitled 'The Granite Controversy'. The present discussion can only present the positions of the protagonists in the most brief of outline form.

In every country where geology was taught as a university degree subject, both of these opposed schools were represented. During the later part of the controversy the Transformists were represented mainly by British, French and Scandinavian geologists, whereas German, Swiss and North American geologists spoke most strongly for the Magmatists. There were, however numerous dissenting voices on both sides, which were very strongly held and argued.

The two different concepts developed slowly during the nineteenth and twentieth centuries, mainly on the basis of field observations made in different places. A new dimension was added however, by the development of experimental petrology in North America and Europe, where geologists attempted to make granite and other rocks by melting powders of appropriate composition in platinum crucibles, and then studying the resulting crystalline products. The chief exponent of this method was N.L. Bowen from the Geophysical Laboratory in Washington who, in a number of papers showed that granite could be produced from basalt by a process of 'crystallisation differentiation'. These results were very influential, and many geologists in every country became adherents of this view, and endeavoured to put it into practice in their own work. However, critics of the crystallisation–differentiation model were able to point out that from a given volume of basaltic magma only a very small proportion of granitic magma would result by that process. Grout (1926) calculated that only about 5% of granitic magma would be produced. This was a serious difficulty for the Magmatists who, however continued to advocate the mechanism. In Britain perhaps the most influential exponent of the concept was S.R. Nockolds (1941) who, in his study of the Garabal Hill Complex in Scotland, considered that the granites of the complex were produced as a result of 'crystallisation–differentiation from earlier basic material'.

There were, however other strands in the debate which appeared from time to time, and which formed the background to the main discussion. One of these was the alternative concept of two primary magmas for the contrasting basaltic and granitic associations. This idea has had a long history and was first proposed by Durocher (1857) and subsequently elaborated by Loewinsson Lessing (1911) who proposed that there were two primordial magmas, one acid and one basic. Many geologists were aware of the concept and were apparently sympathetic to it, including surprisingly enough, both Bowen (1928) and Read (1943). Some of the models proposed for the generation of magmas of different composition from different 'Earth Shells', were surprisingly modern in tone.

In extrusive domains basalts and andesites formed 98% of the total with granites forming only 2%. By contrast in plutonic domains gabbros formed 5% of intrusive rocks whereas granitoids predominated with 95%. This suggested that basalts and granites resulted from different kinds of magmatism in contrast to the one magma origin advocated by Bowen (1928).

These differences were eventually most clearly articulated by Kennedy & Anderson (1938) who suggested that there were two magmatic associations.

1 A volcanic association. Predominantly of basaltic magmas by melting deep levels of the earth's basaltic shell (the mantle in modern terminology).
2 A plutonic association. Predominantly of granitic composition. Despite their deep-seated location they originate at a higher level and are derived from a primary universal granodiorite parent magma.

'The actual mode of irruption differs in the two cases. The granite and granodiorite batholiths penetrate slowly upwards, accompanied by waves of granitisation and migmatisation of the country rock, until arrested by some unknown form of hydrostatic control before reaching the surface. The ascent of basaltic magma is entirely different. No vast inter-crustal reservoirs are formed and the basalt melts appear to be irrupted directly towards the surface through a system of relatively narrow dyke like fissures.' The abundance of basaltic dykes was interpreted as supporting this interpretation.

Although Read (1943) welcomed this concept, since it countered the role of basalt as the universal source, he and others of his persuasion continued to argue for the Transformist concept of granite geology which was essentially non-magmatic in character.

For a period of about 15 years from shortly before the 1939–1945 war until about 1955 a period of intense debate ensued concerning the nature and origin of granitic rocks, in which the old battle lines continued to be represented by the former protagonists, and during which new concepts of granitisation were developed. The Magmatists continued to be represented by Bowen and others of his school who held to their view that basalts were the only primary magma and that other rocks were derived from them by differentiation.

On the Transformist side, the common association of granites with metamorphic rocks continued to be emphasised, and the concept of their origin by some form of transformation of metamorphic rocks into granite via migmatites was developed. This took a number of forms, and one of the concepts which appeared during that period was that the transformation of metamorphic rocks into granite took place 'in the solid' with no liquid phase involved. The language does however tend to get a little involved here and words such as 'ichors' and 'juices' began to be used.

There was a more extreme school of thought, developed principally by the French school (Perrin & Roubalt, 1941) which rejected the concept of even the most attenuated of fluids, and considered that the granitisation of metasedimentary rocks occurred as a result of 'ionic diffusion'. This was thought to be the passage of a wave of the ions of potassium and sodium through the metasedimentary rocks, forming a 'front', driving before it the ions of magnesium, iron and calcium in a mafic front and leaving behind the former metamorphic rock, now reconstituted as a granite. The dioritic margins to some granite bodies were considered to have formed in this way. The protagonists of this concept became known as the 'Dries', while those of less unbending spirit prepared to consider the role of 'ichors', became known as the 'Wets'.

There is some doubt as to whether the Transformists considered it possible for a magma to have formed in a situation of ultrametamorphism and those of the most extreme position probably did not consider it to be possible. Read however, certainly used the term 'migma' which implied at least a certain mobility, and in his final concept of the 'granite series' he advocated a model whereby granites were born in a domain of metamorphites, migmatites and anatectites from which they subsequently moved, to be emplaced at higher levels as plutons in a non-metamorphic domain, where they developed thermal metamorphic aureoles. It seems that Read came to believe that ultrametamorphism ultimately gave rise to granite magmas by the partial melting of metasedimentary rocks.

The controversy over granite origins faded from the forefront of geological interest during the mid-fifties, mainly because the chief protagonists had said everything they could, and there seemed to be no resolution of the problem. The Magmatists still held that granites originated by crystallisation differentiation from basalt, and the Transformists that they were indissolubly related to metamorphic rocks, which gave rise to them by a variety of processes ranging from ultrametamorphism, to granitisation in the solid by ichors or ionic diffusion. The only observations which seemed to offer some prospect of a way forward was the concept of Kennedy & Anderson (1938) that there was a 'volcanic association' and a 'granite association' which they considered to be quite separate, and which was essentially a modernised version of the earlier concepts of Durocher (1857) and Loewisson-Lessing (1911).

Most geologists continued with their own work, keeping their minds as open as possible in the light of the different interpretations which were then available and the influence of their teachers. Pitcher, who was a student of Read, began work on the Donegal Granite in north-west Ireland, on a suite of granites which Read had considered to be exemplary examples of the concept of granitisation. Read (1956) considered some of these granites to be prime examples of granitisation in place because of the phenomenon of 'ghost stratigraphy' which was well preserved in the Donegal Granite. This consisted of the continuation within the granite of trains of inclusions of metamorphic rocks which maintained the same lithostratigraphic position within the granite as they had outside it. These inclusions were mainly of quartzite and marble and it was considered that many of the intervening pelitic members had been granitised in place leaving the quartzites and marbles unaffected as 'resistors'. Pitcher, however was able to demonstrate by careful mapping, that although contacts between the granite and the metasedimentary bands were largely concordant, there were occasional transgressive contacts which indicated the presence of a mobile magma at those situations and which eventually led to the recognition of an intrusive origin for granite emplacement (Pitcher & Berger, 1972).

Pitcher subsequently worked on the Coastal Batholith of Peru which consisted of numerous plutons of mainly tonalitic and granodioritic composition, but with some monzogranites, some of which were in the form of ring complexes. These components were emplaced into host rocks of mainly basaltic composition by processes of magmatic sloping. Work of this nature and elsewhere had begun to cast doubt on the concept of a necessary association of granites with metamorphic rocks, as had the former work of Iddings and Daly.

It was during this period of the mid-fifties and sixties, that rapid advances were made in the fields of chemical analysis, and of isotopic dating, first by the K/Ar method and subsequently by the Rb/Sr method. The application of these techniques were to have a profound and lasting effect on the study of granitic rocks, as well as on many other aspects of geology. Publications of the results of these techniques for granitic rocks began to appear from the mid-sixties onwards, especially from the Sierra Nevada of the western USA, and the use of the Rb/Sr method began to provide information on the kinds of source regions which gave rise to granitic rocks. In particular, variation in chemical and isotopic composition were reported from a profile across the width of the Batholith (Bateman et al., 1963).

Subsequently the results of a similar study of this nature combining petrographic, geochemical and isotopic results from numerous mapped plutons from the Lachlan Fold belt in Australia was reported by Chappell & White (1974). This paper identified two distinct granite types, which were characterised by specific mineralogical, chemical and isotopic features which indicated that their origin was from source regions of different composition. I–types were believed to have had an igneous parentage and S–types a sedimentary one. The groups were distinguished by a number of differences and especially by two geochemical indices, the peraluminous–metaluminous index of Shand (1927, 1947) and the Na_2O/K_2O ratio. Chappell & White, however, did not connect their granite types with different tectonic settings, possibly because in the Lachlan Fold Belt the tectonic setting was the same. It was only the source regions which were believed to be different.

This work regenerated the former controversy, but in a rather different form, because during the intervening years the concept of plate tectonics had become established, and consequently there were new ideas available for the consideration of these results. Within a relatively short time however, the I–S concept had become accepted and integrated in accordance with the rapidly expanding implications of the plate-tectonic model (Pitcher, 1979, 1983). Although the terminology is now commonly used within a plate tectonic framework, that usage has continued to be criticised by some geologists (Clarke, 1992, Atherton, 1993).

Recently Villaseca et al. (1998) have used the peraluminous index of Shand, in combination with the A–B diagram of Debon & le Fort (1983), for a number of different data sets ranging from Island Arcs to the Himalayas. They found that these populations could be reliably distinguished by these parameters on the A–B diagram. They also combined the data with the results of experimental melting studies, and found that the value of the index for the different groups corresponded with the results of partial melting experiments from three main sources, amphibolites, meta-igneous, greywackes and pelites. They emphasised the importance of the source region rather than tectonic setting, as had Chappell and White (1974), but acknowledged that the differences were linked to tectonics, and that the patterns of granite magmatism did seem to be associated with different settings.

With respect to the granite controversy it seems that all the earlier protagonists were partially right and partially wrong. The end result is that all granites were generated by partial melting from protoliths of different composition. These are mainly of basaltic or mantle type in oceanic and continental margin settings and from a range of meta-igneous and metasedimentary types in collisional and continental interior settings.

The origin and association of granites with migmatites which was maintained by the granitisation school, has been confirmed by studies in Brittany (Brown, 1979, in Canada by Sawyer (1996) and in Ecuador by Litherland et al. (1994). However, many of the examples which were formerly advanced as good examples of graniti-sation, have been shown by later studies to have been the result of emplacement processes affecting granites of crustal origin, whether of I–type or S–type.

The concept of fractional crystallisation and differentiation is universally accepted for every kind of igneous rock and Harker diagrams are regularly used in all publications.

The new concept of different source regions for different kinds of granite now prevails. It is considered that most granites result from the partial or complete melting of crustal source regions having a range of composition from wholly igneous (basaltic) to wholly sedimentary (pelitic). In some cases the mantle is also thought to play a role. Whether these source regions are systematically linked to different tectonic settings remains to be established. In general this does seem to be the case, but there are a number of exceptions which advise caution in applying a generalised model, and study of these may result in useful new insights.

In retrospect we can see that many of the concepts that we now accept as being the result of plate-tectonic processes had been foreshadowed in embryo by earlier workers. The concept of two separate magmas for basaltic and granitic rocks respectively underwent a long history of sporadic development from Durocher (1857) via Loewinsson-Lessing (1911) through Iddings (1909) and Daly (1911) who suggested that there might be two granite series. The later work of Kennedy and Anderson (1938) which identified a Volcanic Association and a Plutonic Association resulting from different levels of magma genesis, provided a model for granite genesis which was only fully realised through the later developments in tectonic theory, and the simultaneous advances in the techniques of geochemistry and isotopic dating.

The former 'Granite Controversy' is now only of historic interest. Granites continue to be controversial, but in ways that differ from the concepts of the former protagonists. Many of the observations, insights, and even prejudices which were debated during the controversy, have now been shown to have been reflections of real differences in granite geology, which have now been welded together into an integrated body of knowledge, and which provides a sound basis for the pursuit of granite studies, and especially for the theme of this work, namely, that of mapping granite batholiths.

9 Epilogue

In writing this booklet I have indicated that in spite of the great diversity of individual granite bodies, they are all systematically linked by common processes developed in different kinds of source regions by a range of tectonic mechanisms within the plate-tectonic framework of the Earth generated by mantle convection. The mechanisms which affect oceanic and marginal domains, result in granitoids with relatively straightforward geological characteristics, but in those many cases where the source region is continental crust of varying age and complexity of composition, the relationship of granite type to a specific tectonic process can be a matter of of difficulty or uncertainty. In spite of this it remains true that the connection between tectonics and granite generation is real and that the granites we see reflect that connection through time and space. While the main areas of conformity of granitoid with source and tectonic setting are reasonably well defined, the areas of uncertainty increase with increasing complexity of source region and the driving tectonic forces. Because of this there are certain areas of granite geology which are imperfectly understood within the parameters of present knowledge. These are still severely constrained by the lack of sufficiently detailed knowledge of the geology of granite belts and batholiths, especially in the domain of predominantly crustal granites. This area needs to be systematically addressed and those geologists lucky enough to be working in such regions can look forward to an interesting time.

This view, as simply stated above, has only recently come to be generally accepted by the geological community, and is one of the many areas of geology to have become more or less unified by the theory of plate tectonics. Prior to the development of that theory granites were studied by geologists mainly within the parameters of their local and regional geology which were of course, extremely varied and which, among other things, gave rise to the granite controversy outlined in the former chapter.

I was a student at that time and among my teachers were those who favoured either side of the controversy. It was certainly a liberal education, and among other things I was exposed to the work of Kennedy and his concepts of basaltic magma types and of a volcanic and plutonic association. These ideas were certainly relevant to the controversy, but were largely overlooked, even though the contemporary work of Holmes (1944) on global tectonics did provide an appropriate framework for their further development.

In my own case I eventually graduated and subsequently worked on the metamorphic and granitic rocks of the Caledonides in Ireland. Perhaps the situation then

J. Cobbing: LNES 96, pp. 117–121, 2000.

was much as Read (1956) had said 'that every twenty years or so, the problem has been finally settled and a sort of uneasy peace has broken out'.

In the mid-sixties I was posted to a geological mission in Peru as a field geologist, and was given the task of mapping several quadrangles, which included large areas of the Coastal Batholith. It did not take long to find that the Batholith had a compositional range from gabbro to granite, and that the most abundant rock types were tonalite and granodiorite. The envelope consisted of volcanic rocks of Cretaceous age, only a little older than the Batholith itself, and there were no metamorphic rocks. Even metamorphic aureoles were restricted to a metre or so or, more commonly, were not visible at all.

I became involved with Wallace Pitcher and his group, who coincidentally began work on the Batholith at the same time. In their case it was by choice, to gain experience of unfamiliar granites, whereas in mine it was by chance. However we worked fruitfully together and in 1973 we attended a meeting of the Circum Pacific Plutonism Group in Chile. The Group was organised by Paul Bateman of the USGS and the meeting was hosted by geologists of the University of Santiago led by Lucho Aguirre. It consisted of a couple of days of talks followed by several days of field excursions in Chile and Argentina.

In one of the talks, the concept of I and S–type granites was floated by Alan White. There was a fairly robust discussion in which the main contributors were Wallace Pitcher and Paul Bateman. On the whole the reception was thoughtful and interested, but fairly low key, and as can be imagined, gave rise to considerable discussion during the subsequent excursions. Most of us only became fully aware of the implications of the concept when the paper was published in the proceedings of the meeting (Chappell, B.W. & White, A.J.R, 1974).

It is difficult to say in retrospect what my own views were on this question, but I remember writing a visit report which said 'there might be something in it' which I think fairly reflected the opinions of most of us. In effect the concept has replaced the older 'granite controversy' and some people might think it has simply restated it in different form. By linking the different 'granite types' to different source regions of different composition, Chappell and White both resurrected the debate and transformed it, and it has been the dominant theme of granite geology for the last few decades.

Certainly with respect to S–types the link with a metasedimentary source material has now been demonstrated in a number of cases. (Barbero & Villaseca, 1992, Brown, 1979, Litherland et al., 1994 and Sawyer, 1992 have described granites, in association with migmatites, segregating and forming plutons. These studies have demonstrated that the generation of granite of this sort does actually occur in high grade metamorphic terrains, and that these small scale occurrences have been shown to coalesce into plutons which are emplaced to higher level. The Cancale pluton in Brittany is one such example (Brown, 1979). It has also been found that S–type granites with abundant relics of a metamorphic protolith can be of batholithic dimensions (Litherland et al., 1994). Examples such as these show that the earlier concept of Read is correct for granites of this type. However, for the various categories of I–type granitoids such a direct connection with a source region cannot be satisfactorily demonstrated by classical field methods.

It would seem that I–types, in general were generated at deeper crustal levels and become separated from their source regions to a greater degree than is the case with some S–types. Consequently it is only possible to infer the nature of the possible source region by geochemical and isotopic studies. Fortunately such methods of investigation are now widespread and have contributed to the resolution of this question. Common source regions for I–type granitoids are the basaltic lower crust or the mantle wedge in subduction zones.

It is only the most recent generation of geologists which has had the good fortune of having had access to the most modern methods of geological investigation in anything approaching a routine way. The use of isotopic, geochemical and geophysical methods has mushroomed over the past 30 years. Prior to that geologists had to rely for the most part on the methods of classical geology which, in the words of one authority, was "to gather every possible clue, climb out on a limb, and guess" (Ardrey, 1961). Many of these guesses have been confirmed by later, more rigorous methods, but others have now passed into the realm of geological folklore.

How divergent these guesses could be, is illustrated by the conceptual chasm separating those who believed in continental drift, and those who denied it. The original concept proposed by Wegener (1924) was based on the apparent fit of the opposing shores of Africa and South America, together with certain geological similarities on either side of the Atlantic Ocean. These were systematically elaborated by DuToit (1937) who assembled different strands of geological evidence such as the dispersal of Gondwana glacial deposits, coal basins, and the location of plateau basalts in support of the model. This evidence was dismissed by the opponents of the hypothesis, some of whom constructed completely hypothetical land bridges across the oceans, in order to explain the occurrence of identical fossil faunas of Lower Palaeozoic age, in widely separated continents, and others who made mathematical calculations which 'proved' that no conceivable mechanism was adequate to generate the proposed separation.

Other geologists were generally more favourably disposed towards the concept of drift. Indeed Holmes (1944) in his influential work for first-year geology students, even proposed the generation of new oceanic crust as the result of mantle convection. His model showed rising mantle plumes situated along the mid-ocean ridges producing new basaltic oceanic crust, with the older oceanic crust descending beneath the island arcs and continents of the Pacific rim to produce orogenic belts. It seems astonishing now that the concept of plate tectonics did not develop from that basis, but had to wait a further quarter of a century for the geophysical mapping of the ocean floor and the recognition of normally and reversely magnetised basaltic eruptions, for the implications of sea floor spreading to be fully appreciated.

We all now know that geophysical surveys of the ocean floor led to the identification of sea floor spreading and the theory of plate tectonics. An ultimate victory for the empiricists. The lesson for all of us is, I think, to respect the geological evidence whatever form it may take.

The concept of plate tectonics has so far proved to be very robust, and virtually every aspect of geology is now routinely viewed within that light. It does not make sense to exclude granites, since they are but one of many processes brought into action and following a sequence of rhythmic development within a wider framework of plate convergence or extension. For these reasons I am happy to follow Pitcher (1978, 1983) who first adapted and expanded the I–S notation of Chappell & White to conform with the gross patterns of plate convergence. Although these patterns are essentially simple, they can be extremely complex. It is possible to find S–type granites in Andean regions and there are many other examples of apparently anomalous occurrences. However, in most cases there is generally a rational explanation for the occurrences. It is not actually obligatory for geologists to use the I–S notation. We have not yet brought in the thought police. Those, who for whatever reason may be unhappy with the concept, are at perfect liberty to use the tried and tested systems of classical geology which now commonly provide a very complete picture of the nature and origin of most granites.

Certain it is I am sure that plate tectonics is here to stay. Within that concept the I–S expanded system will either survive or not. In the meantime every geologist should be able to follow their instinct or prejudice, having regard for the data.

Although plate tectonics is now well established as the main engine for geological activity of our planet it seems that it was not always so. For most of the Proterozoic it appears to have been in abeyance, or to have been only sporadically active. From between 2.0 and 1.0 Ga the Earth underwent a prolonged period of stability in crustal dynamics accompanied by stability of climate. Several geologists, and most recently Brasier & Lindsay (1998) have suggested that it was this lengthy period of calm which enabled the development of the eukaryotic cell via the symbiosis of the various prokaryotic elements in the Proterozoic oceans, and hence to the biological explosion at the beginning of the Cambrian. It is not at present understood why the Earth should have been so static during that period, but for granite geologists it is interesting to find that much of the igneous activity during that time consisted of rapakivi granites and layered anorthosites, rather than the volcanic arc associations with which most of us are familiar.

Few of us are of the intellectual calibre of the participants in the granite controversy and we should not be too frightened of making mistakes. Any field geologist is bound to make misjudgements at some time or other. This is the occupational hazard of our profession. It happens because of the complexity of the data itself, which may be inadequate to form a definitive judgement, and in some cases may actually be misleading. For example not many people could have predicted from the field evidence that the metamorphic infrastructure of core complexes would actually be younger than the cover. This may be rather an extreme example, but the field evidence can often be very difficult to interpret correctly, yet judgements have to be made, however qualified they might be.

When on field work, geologists of all disciplines operate within the parameters of classical geology, which, in spite of the comments of Ardrey (1961) quoted above, have generally served us pretty well. Most of this book has been devoted to outlining these methods as they relate to the study of granites in the realm of their natural occurrence. Cumulative field observations build up a framework of knowledge which progressively constrains areas of uncertainty. Nowadays it is possible to tackle many of these areas because the appropriate technology is available, even though the necessary human and financial resources to employ them are often in short supply, since granites do not generally rank high on the scale of human priorities.

I think it is probably impossible for a single individual to know everything about granite geology, and the numerous processes which affect granites during their generation and emplacement. These have to be reconstructed by detective work based on field and laboratory studies, in combination with experimental work designed to recreate the criteria observed. This is a pretty tall order considering the complexity of independently evolving factors involved in the formation of even the simplest granite.

I have probably seen and mapped more granites than most geologists, but there are whole areas of granite geology of which I have no direct experience. Until quite recently I had never seen a cordierite-bearing granite, a deficiency which has now been remedied by the kindness of my Spanish friends. Moreover I have never worked with alkaline and anorogenic granites. I suspect that most geologists are similarly limited in this respect. Nevertheless the corporate body of existing knowledge provides a platform for the narrowing of the remaining gaps in our knowledge and the search for a fully comprehensive theory of granite geology.

It may not be such an exalted quest as the search for a unified theory of the universe but it is sufficiently challenging to provide a worthwhile objective for at least another generation of geologists.

10 References

Akaad MK (1956) The Ardara granitic diapir of County Donegal, Ireland. Quarterly Journal of the Geological Society of London 112: 263–288

Angus NS (1962) Ocellar hybrids from the Tyrone Igneous Series, Ireland. Geological Magazine, London 99: 9–26

Angus NS (1971) Comments on the origin of ocellar hybrids. Lithos, Oslo 4: 381–388

Ardrey R (1961) African Genesis. Collins, pp 416

Aspden JA, Fortey N, Litherland M, Viteri F & Harrison SM (1993) Regional S-type granites in the Ecuadorian Andes: possible remnants of the breakup of Gondwana. Journal of South American Earth Sciences 6: 123–132

Aspden JA, Bonilla W & Duque P (1995) The El Oro metamorphic Complex, Ecuador: geology and economic mineral deposits. Overseas Geology and Mineral Resources 67, pp 63

Atherton MP (1990) The Coastal Batholith of Peru: The product of rapid recycling of 'New Crust' formed within a rifted continental margin. Geological Journal 25: 337–349

Atherton MP (1993) Granite magmatism. Journal of the Geological Society of London 150: 1009–1023

Atherton MP, McCourt WJ, Sanderson LM & Taylor WP (1979) The geochemical character of the segmented Peruvian Coastal Batholith and associated volcanics. Origin of Granite Batholiths. Atherton MP & Tarney J (eds) Shiva Press: 45–46

Atherton MP & Sanderson LM (1987) The Cordillera Blanca Batholith: a study of granite intrusions and the relation of crustal thickening to peraluminosity. Geologische Rundschau 76: 213–232

Atherton MP & Petford N (1993) Generation of sodium-rich magmas from newly underplated basaltic crust. Nature 362: 144–146

Atkin BP, Injoque–Espinosa J & Harvey PK (1985) Cu–Fe amphibole mineralisation in the Arequipa segment. In "Magmatism at a Plate Edge", the Peruvian Andes (eds) Pitcher WS, Atherton MP, Cobbing EJ & Beckinsale RD. Blackie Halsted Press, pp 328: 261–270

Avila–Salinas WA (1990) Tin–bearing granites from the Cordillera Real, Bolivia; A petrological and geochemical review. Plutonism from Antarctica to Alaska (eds) Kay SM & Rapela CW. Geological Society of America Special Paper 241: 145–159

Bagnold RA (1954) Experiments on a gravity free dispersion of large solid spheres in a Newtonian fluid under shear. Proceedings of the Royal Society of London 225: 49–63

Barbarin B (1990) Granitoids: main petrogenetic classifications in relation to origin and tectonic setting. Geological Journal 25: 227–238

Barbero L & Villaseca C (1992) The Layos Granite, Hercynian Complex of Toledo (Spain): an example of parautocthonous restite–rich granite in a granulitic area. Transactions of the Royal Society of Edinburgh: Earth Sciences 83: 127–138

Barr SM (1990) Granitoid rocks and Terrane characterisation: an example from the northern Appalachian orogen. Geological Journal 25: 227–238

Barriere M (1977) Deformation associated with the Ploumanac'h intrusive complex, Brittanny. Journal of the Geological Society of London 104: 461–476

Batchelor RA & Bowden P (1985) Petrogenetic interpretation of granitoid rock series using multicationic parameters. Chemical Geology 48: 43–55

Bateman PC, Clarke LD, Huber NK, Moore JG & Rinehart CD (1963) The Sierra Nevada Batholith: a synthesis of recent work across the central part. US. Geological Survey Professional Paper No 414D: 1–45

Bateman PC (1992) Plutonism in the central part of the Sierra Nevada, US. Geological Survey Professional Paper 1483: pp 186

Beckinsale RD (1979) Granite magmatism in the tin belt of Southeast Asia. In origin of granite batholiths: Geochemical evidence (eds) Atherton MP & Tarney J Shiva Publishing 34–44

Berthe D, Choukroune P & Jegouzo P (1979) Orthogneiss, mylonite and noncoaxial deformation of granites: the example of the South Armorican Shear Zone. Journal of Structural Geology 1: 31–42

Bignell JD & Snelling NJ (1977) The geochronology of Malayan granites. Overseas Geology and Mineral Resources 47: 1–72

Bishop AC (1963) Dark margins at igneous contacts. Proceedings of the Geological Association of London 74: 289–300

Blake DH, Elwell RWD, Gibson IL, Skelhorn RR & Walker GPL (1965) Some relations resulting from the intimate association of acid and basic magmas. Quarterly Journal of the Geological Society of London 121: 31–49

Bowden P, Batchelor RA, Chappell BW, Didier J & Lameyre J (1984) Petrological geochemical and source criteria for the classification of granitic rocks: a discussion. Physics of the Earth and Planetary interiors 35: 1 11

Bowen NL (1928) The evolution of the Igneous Rocks. Princeton University Press, Princeton NJ: 334

Brasier MD & Lindsay JF (1998) A billion years of environmental stability and the emergence of the eukaryotes: New data from northern Australia. Geology 28: 555–558

Brown GC, Thorpe RS & Webb PC (1984) The geochemical characteristics of granitoids in contrasting arcs and comments on magma sources. Journal of the Geological Society of London 141: 413–426

Brown M (1973) Definition of metatexis, diatexis and migmatite. Proceedings Geological Association 84: 371–382

Brown M (1979) The petrogenesis of the St Malo Migmatite Belt, Armorican Massif, France, with particular reference to the diatexites. Neues Jahrbuch fur Mineralogie Abhandlungen, 135: 48–74

Brown M (1994) The generation, segregation, ascent and emplacement of granite magma. The migmatite to crustally derived granite connection in thickened orogens. Earth Science Revue 36: 83–130

Brown M, Power GM, Topley CG & D'Lemos RS Cadomian magmatism in the North Armorican Massif. In D'Lemos RS, Strachan RA & Topley CG (eds) 1990. The Cadomian Orogeny. Geological Society of London Special Publication 51: 81–213

Brun JP, Gapais D, Cogne JP, Ledru P & Vigneresse JL (1990) The Flamanville Granite (North West France): an unequivocal example of a tectonically expanding pluton. Geological Journal 25: 271–286

Bussell MA (1985) The centred complex of the Rio Huaura: a study of magma mixing and differentiation in high level magma chambers. In "Magmatism at a Plate Edge" the Peruvian Andes (eds) Pitcher WS, Atherton MP, Cobbing EJ & Beckinsale RD Blackie Halsted Press, pp 328: 128–153

Castro A (1986) Structural patterns and ascent model in the Central Extremadura Batholith, Hercynian Belt, Spain. Journal of Structural Geology 8: 633–645

Chappell BW (1984) Source rocks of S and I-type in the Lachlan Fold Belt, Southeast Australia. Philosophical Transactions of the Royal Society of London A320: 693–707

Chappell BW (1996) Compositional variation within granite suites of the Lachlan Fold Belt: its causes and implications for the physical state of granite magma. Transactions of the Royal Society of Edinburgh 88: 159–170

Chappell BW & White AJR (1974) Two contrasting granite types. Pacific Geology 8: 173–174

Chappell BW & White AJR (1989) Some supracrustal (S–type) granites of the Lachlan fold belt: in The origin of granites. Transactions of the Royal society of Edinburgh; Earth Sciences 79: 169–181

Chappell BW & White AJR (1991) Restite enclaves and the restite model. In Enclaves and granite petrology (eds) Didier J & Barbarin B Elsevier, pp 601: 375–381

Chappell BW & White AJR (1992) I & S–type granites in the Lachlan Fold Belt. Transactions of the Royal Society of Edinburgh: Earth Sciences 83: 1–26

Chappell BW, White AJR & Wyborn D (1977) The importance of residual source material (restite) in granite petrogenesis. Journal of Petrology 18: 1111–1138

Chappell BW & Stephens WE (1988) Origin of infracrustal I–type granite magmas. Transactions of the Royal Society of Edinburgh: Earth Sciences 79: 71–86

Chappell BW, White AJR & Hine R (1988) Granites and Basement terranes in the Lachlan Fold Belt southeastern Australia. Australian Journal of Earth Sciences 35: 515–521

Clarke DB (1992) Granitoid Rocks. Topics in Earth Sciences 7. Chapman & Hall, New York, pp 296

Cloos E (1936) Der Sierra Nevada Pluton in California. Neues Jarbuch fur Mineralogie, geologie und Paleontologie, Abhandlungen, sec B, 76: 355–450

Cloos H (1941) Bau und Tatigkeit von Tuffschloten: Untersuchungen An Dem Schwabischen Vulkan. Geologische Rundschau 32: 709–800

Cluzel D, Lee BJ & Cadet JP (1991) Indosinian dextral ductile fault system and synkinematic plutonism in the southeast of the Ogcheon Belt (South Korea). Tectonophysics 194: 131–151

Cobbing EJ (1982) The segmented Coastal Batholith of Peru; Its relationship to volcanicity and metallogenesis. Earth Science Reviews 18: 241–251

Cobbing EJ (1990) A comparison of granites and their tectonic settings from the South American Andes and the Southeast Asian Tin Belt. In Kay SM & Rapela CW (eds) Plutonism from Antarctica to Alaska. Geological Society of America Special Paper 241: 193–204

Cobbing EJ, Manning PI & Grffith AE (1965) Ordovician–Dalradian unconformity in Tyrone. Nature 206: 1132–1135

Cobbing EJ & Pitcher WS (1972) the Coastal Batholith of Central Peru. Journal of the Geological Society of London 128: 421–450

Cobbing EJ, Pitcher WS & Taylor W (1977) Segments and Super–units in the Coastal Batholith of Peru. Journal of Geology. Chicago 85: 625–631

Cobbing EJ, Mallick DIJ, Pitfield PEJ & Teoh LH (1986) The granites of the Southeast Asian Tin Belt. Journal of the Geological Society of London 143: 537–550

Cobbing EJ, Pitfield PEJ, Darbyshire DPF & Mallick DIJ (1992) The granites of the Southeast Asian Tin Belt. British Geological Survey Overseas Memoir 10. HMSO: pp 369

Coney PJ (1980) Cordilleran metamorphic core complexes: an overview. In MD Crittenden et al (eds) Geological Society of America Memoir 193: 1–25

Coney PJ (1992) The Lachlan Fold Belt of Australia and Circum Pacific tectonic evolution. Tectonophysics 214: 1–25

Craxton CW (1968) Mineral layering in the Galway Granite, Connemara, Eire. Geological Magazine 105: 149–159

Debon F & Le Fort P (1983) A chemical–mineralogical classification of common plutonic rocks and associations. Transactions of the Royal Society of Edinburgh: Earth Sciences 73: 135–149

Didier J (1973) Granites and their enclaves, The bearing of enclaves on the origin of granites. Developments in Petrology 3 Elsevier, pp 393

Didier J (1987) Contribution of enclave studies to the understanding of origin and evolution of granitic magmas. Geologische Rundschau 76: 41–50

Didier J & Barbarin B (1991) Enclaves and granite petrology. Elsevier, pp 601

Didier J, Duthou JL & Lameyre J (1982) Mantle and crustal granites: genetic classification of orogenic granites and the nature of their enclaves. Journal of Volcanological and Geothermal Research 14: 125–132

Durocher J (1847) Note sur une espèce de granite provenant de la Normandie et de la Bretagne. Bull. Geol Soc France 4: 140

Du Toit AL (1937) Our Wandering Continents. Oliver and Boyd, Edinburgh, pp 366

Elburg MA & Nicholls IA (1995) Origin of microgranitoid enclaves in the S–type Wilson's Promontory Batholith, Victoria: evidence for magma mingling. Australian Journal of Earth Sciences 42: 423–435

Emeleus CH (1963) Structural and petrological observations on layered granites from southern Greenland. Mineralogical Society of America, Special Paper 1: 22–29

Eskola P (1948) The problem of mantled gneiss domes. Quarterly Journal of the Geological Society of London 104: 461–476

Gastil RG, Phillips RP & Allison EC (1975) Reconnaissance geology of the state of Baja California. Memoir Geological Society of America 140, pp 170

Gould SJ (1989) Wonderful Life: The Burgess Shale And The Nature Of History. Hutchinson, p 347

Grout FF (1926) The use of calculations in petrology. Journal of Geology 34: 549

Harker A (1904) Memoirs of the Geological Survey of Scotland, pp 481

Harry WT & Emeleus CH (1960) Mineral layering in some granite intrusions of SW Greenland. International Geological Congress 21st session 4: 172–181

Harris NBW, Pearce JA & Tindle AG (1986) Geochemical characteristics of collision–zone magmatism. In Collision Tectonics. Coward MP & Ries AC (eds) Geological Society Of London Special Bulletin 19: 67–81

Hatch FH, Wells AK & Wells MK (1951) Petrology of the igneous rocks. Thomas Murby London, pp 515

Haughton S (1862) Experimental researches on the granites of Ireland: Part 3. On the granites of Donegal. Quarterly Journal of the Geological Society of London 18: 403–420

Henley S (1974) Geochemistry and petrogenesis of elvan dykes in the Perranporth area, Cornwall. Proceedings of the Ussher Society 3: 136–137

Heltzel R, Ring V, Akal C & Troesch M (1995) Miocene NNE directed extensional unroofing in the Menderes Massif southwestern Turkey. Journal of the Geological Society of London 152: 639–654

Hill EJ, Baldwin SL & Lister GS (1992) Unroofing of active metamorphic core complexes in the D'Entrecasteaux Islands, Papua New Guinea. Geology 20: 907–910

Holder MT (1979) An emplacement mechanism for post tectonic granites and the implications for their geochemical features. In Origin of Granite Batholiths (eds) Atherton MP & Tarney J. Shiva Press: 116–128

Hubbard MG, Spencer DA & West DP (1995) Tectonic exhumation of the Nanga Parbat Massif, northern Pakistan. Earth and Planetary Science Letters 133: 213–225

Hudson T & Arth JG (1983) Granites of the Seward Peninsula, Alaska. Geological Society of American Bulletin 94: 768–790

Hutton DW (1982) A tectonic model for the emplacement of the Donegal Granite, NW Ireland. Journal of the Geological Society of London 139: 615–631

Hutton DW (1988) Granite emplacement mechanisms and tectonic controls: inferences from deformation studies. Transactions of the Royal Society of Edinburgh: Earth Sciences 83: 377–382

Hutton DW (1992) Granite sheeted complexes: evidence for dykeing ascent mechanism. Transactions of the Royal Society of Edinburgh: Earth Sciences 83: 377–382

Hutton DW, Aftalion M & Halliday AN (1985) An Ordovician ophiolite in County Tyrone, Ireland. Nature 315: 210–212

Hutton DW & Reavy RJ (1992) Strike–Slip tectonics and granite petrogenesis. Tectonics 5: 960–967

Hutton DW & Ingram GM (1992) The Great Tonalite Sill of south-eastern Alaska and British Columbia: emplacement into an active contractional high angle reverse shear zone. Transactions of the Royal Society of Edinburgh: Earth Sciences 83: 381–386

Hyndman DW (1983) The Idaho Batholith and associated plutons. Idaho and western Montana. In Roddick AJ (ed) Cicum Pacific plutonic Terranes. Geological Society of America Memoir 159: 213–240

Irvine TN & Barager WRA (1971) A guide to the chemical classification of the common volcanic rocks. Canadian Journal of Earth Sciences 8: 523–548

Iddings JP (1909) Igneous Rocks. John Wiley & Sons, New York

Ishihara S (1977) The magnetite series and ilmenite series granitic rocks. Mining Geology 27: 293–205

Jegouzo P (1980) The South Armorican Shear Zone. Journal of Structural Geology 2: 39–47

Johannsen A (1937) A descriptive petrography of the igneous rocks. University of Chicago Press, 4 vols

Kay SH, Kay RW, Citron GP & Perfit MR (1992) Calc–alkaline plutonism in the intra–oceanic Aleutian arc, Alaska. In Kay SM & Rapela CA (eds) Plutonism from Antarctic to Alaska. Geological Society of America Special Paper 241: 233–255

Kelly J (1853) On the quartz rocks of the northern part of the County of Wicklow. Journal of the Geological Society of Dublin 5: 237–276

Kennedy WQ (1933) Trends of differentiation in basaltic magmas. American Journal of Science: 239–256

Kennedy WQ & Anderson EM (1938) Crustal Layers and the origin of magmas. Bulletin of Vulcanology: series ii 3–24

Kistler RW & Peterman ZE (1978) Reconstruction of crustal blocks of California on the basis of initial strontium isotopic compositions of Mesozoic granitic rocks. U.S. Geological Survey Professional Paper 1071: pp 17

Kontak DJ, Clarke AH, Farrar E & Strong DF (1985) The rift associated Permo–Triassic magmatism of the Eastern Cordillera; a precursor to the Andean orogeny. In Magmatism at a Plate Edge. The Peruvian Andes. (eds) Pitcher WS, Atherton MP, Cobbing EJ & Beckinsale RD. Blackie Halsted Press: pp 328

Kuno H (1969) Andesite in time and space. Bulletin of the Oregon Department of Geology and Mineral Industry 65: 13–20

Lameyre J (1983) Le plutonisme océanique intraplaque: Exemple des isles Kerguelen. Comité Nacional Francais des Recherches Antarctiques 54: pp 290

Lameyre J & Bowden P (1982) Plutonic rock type series: discrimination of various granitoid series and related rocks. Journal of Volcanological and Geothermal Research 14: 169–186

Lameyre J & Bonin B (1991) Granites in the main plutonic series. In Enclaves and granite petrology. (eds) Didier J & Barbarin B. Elsevier, pp 601

La Roche H de (1978) La chimie des roches presentee et interprétee d' après la structure de leur facies mineral dans l'espace des variables chimiques: fonctions specifique et diagrammes qui s'en déduisant. Application aux roches ignées. Chemical Geology 21: 87–93

Leake BE (1990) Granite magmas: their sources, initiation and consequences of emplacement. Journal of the Geological Society of London 147: 570–589

Lehmann B (1987) Tin granites, geochemical heritage, magmatic differentiation. Geologische Rundschau 76: 177–185

Lehmann B (1990) Metallogeny of tin. Lecture notes in Earth Sciences. Springer Verlag Berlin, pp 211

Liew TC & McCulloch, MT (1985) Genesis of granitoid batholiths of Peninsular Malaysia and implications for models of crustal evolution: Evidence from a Nd-Sr isotopic and U-Pb zircon study. Geochemica et Cosmochimica Acta 49: 587–600

Lister GS, Banga G & Fenestra A (1984) Metamorphic core complexes of Cordilleran type in the Cyclades, Aegean Sea, Greece. Geology 12: 221–225

Litherland M, Aspden JA & Jemelieta RA (1994) The Metamorphic Belts of Ecuador. Overseas Memoir of the British Geological Survey No 11, pp 147

Loewinsson-Lessing (1911) The fundamental problems of petrogenesis, or the origin of the igneous rocks. Geological Magazine 8: 248

Loiselle MC & Wones DR (1979) Characteristics of anorogenic granites. Geological Society of America, Abstracts with Programmes 11: 468

MacDonald AS, Barr SM, Dunning GR & Yaoanoiythin W (1993) The Doi Ithanon metamorphic core complex in NW Thailand: age and tectonic significance. Journal of Southeast Asian Earth Sciences 8: 117–125

Michel-Levy A (1877) L'origine des terrains cristallins primitifs. Bulletin Societé Geologique France 3ser 16: 102

Miller CF & Barton MD (1990) Phanerozoic plutonism in the Cordilleran Interior, U.S.A. In Kay SM & Rapela CW (eds) Plutonism from Antarctic to Alaska. Geological Society of America Special Paper 241: 213–231

Miller CF, Hanchar JM, Wooden JL, Bennett VC, Harrison TM, Wark DA & Foster DA (1992) Source region of granite batholith: evidence from lower crustal xenoliths and inherited accessory minerals. Transactions of the Royal Society of Edinburgh: Earth Sciences 83: 49–62

Mitchell AHG & Carlile JC (1994) Mineralisation, antiforms and crustal extension in andesitic arcs. Geological Magazine 131: 231–242

Mukasa SB & Tilton GR (1985) Zircon U–Pb ages of super–units in the Coastal Batholith of Peru. In "Magmatism at a Plate Edge" The Peruvian Andes (eds) Pitcher WS, Atherton MP, Cobbing EJ & Beckinsale RD. Blackie Halsted Press, pp 328

Mukasa SB & Henry DJ (1990) The San Nicolas batholith: evidence for an early Palaeozoic magmatic arc along the continental margin of Peru. Journal of the Geological Society of London 147: 27–39

Myers JS (1974) Cauldron subsidence and fluidisation: mechanism of intrusion of the Coastal Batholith of Peru into its own ejecta. Bulletin of the Geological Society of America 86: 1209–1220

Nachit H, Razafimahefa N, Stussi JM & Carron JP (1985) Composition chimiques des biotites et typologie magmatique des granitoids. Compte Rendus de L'Academie des Sciences. Paris 301: 813–818

Nockolds SR (1941) The Garabal Hill-Glen Fyne Igneous Complex. Quarterly Journal of the Geological Society of London 96: 451–510

Nordgulen O, Bickford ME, Nissen KL & Wortman GL (1993) U–Pb ages from the Bindal Batholith and the tectonic history of the Helgeland Nappe Complex, Scandinavian Caledonides. Journal of the Geological Society of London 150: 771–783

Parada MA (1990) Granitoid plutonism in central Chile and its geodynamic implications; a review. In Kay S & Rapela CW (eds) Geological Society of America Special Paper 241: 349–363

Paterson SR, Vernon RH & Tobisch OT (1989) A review of criteria for the identification of magmatic and tectonic foliations in granitoids. Journal of Structural Geology 11: 349–363

Peacock MA (1931) Classification of igneous rocks. Journal of Geology 39: 54–67

Pearce JA & Cann JR (1973) Tectonic setting of basic igneous rocks investigated using trace element analysis. Earth and Planetary Science Letters 19: 290–300

Pearce JA, Harris NBW & Tindle AG (1984) Trace element discrimination diagrams for the interpretation of granitic rocks. Journal of Petrology 25: 956–983

Peccerillo A & Taylor SR (1976) Geochemistry of Eocene Calc–Alkaline volcanic rocks from the Kastamonu area, Northern Turkey. Contributions to Mineralogy and Petrology 58: 63–81

Perrin R & Roubalt M (1941) Observation d'un 'front' de metamorphisme regional. Bulletin Societé Geologique de France 5: 11–183

Petford N & Atherton MP (1992) Granitoid emplacement and deformation along a major crustal lineament: the Cordillera Blanca, Peru. Tectonophysics 205: 171–185

Petford N, Kerr RC & Lister JR (1993) Dike transport of granitoid magmas. Geology 21: 845–848

Pichowiak S, Buchelt M & Damm KW (1990) Magmatic activity and tectonic setting of the early stages of the Andean Cycle in northern Chile. In Kay S & Rapela CW (eds) Geological

Society of America Special Paper 241, Plutonism from Antarctica to Alaska 127–144

Pitcher WS (1969) In Newall G & Rast N (eds) Mechanism of Igneous Intrusion. Geological Journal Special Publication No 2: 123–140

Pitcher WS (1978) Comments on the geological environment. In Origin of granite batholiths: geochemical evidence (eds) Atherton MP & Tarney J. Shiva Press, pp 148: 1–8

Pitcher WS (1979) The nature, ascent and origin of granite magmas. Journal of the Geological Society of London 136: 627–662

Pitcher WS (1983) Granite type and tectonic environment. In Hsu K (ed) Mountain building processes. Academic Press, London: 19–45

Pitcher WS (1993) The nature and origin of granite. Chapman Hall, pp 321

Pitcher WS & Berger AR (1972) The Geology of the Donegal Granite; a study of granite emplacement and unroofing. John Wiley & Sons. New York, pp 435

Pitcher WS & Bussell MA (1977) Structural control of batholith emplacement in Peru: a review. Journal of the Geological Society of London 133: 249–256

Pitfield PIJ, Teoh LH & Cobbing EJ (1990) Textural variation and tin mineralisation in granites from the Main Range Province of the Southeast Asian Tin Belt. Geological Journal 25: 419–429

Poldevaart A & Taubeneck WH (1959) Layered intrusions of Willow Lake type. Bulletin of the Geological Society of America 70: 1395–1398

Pupin JP (1980) Zircon and granite petrology. Contributions to Mineralogy and Petrology 73: 207–220

Rapela CW, Toselli A & Saavedra J (1990) Granite plutonism in the Sierra Pampeanas; an inner Cordilleran Palaeozoic arc in the southern Andes. In Kay S & Rapela CW (eds) Geological Society of America Special Paper 241: 77–90

Read HH (1943) Meditations on Granite. Part One. Proceedings of the Geological Association 54: 64–85

Read HH (1944) Meditations on Granite Part Two. Proceedings of the Geological Association 55: 44–93

Read HH (1956) The granite controversy: Geological addresses illustrating the evolution of a disputant. Thomas Murby, pp 430

Roberts RP & Clemens JD (1993) Origin of high potassium I–type granitoids. Geology 21: 825–828

Roddick JC & Armstrong JE (1959) Relict dykes in the Coast Mountains near Vancouver B.C. Journal of Geology 67: 603–613

Rongfu Pei & Dawei Hong (1995) The granites of south China and their metallogeny. Episodes 18: 77–82

Rosenbusch H (1887) Microscopische, Physiographie der Mineralien und Gesteine. Massige Gesteine, Stuttgart

Rosenbusch H (1877) Die stieger schiefer unf ihre contact zone en dan granititen von Barr-Andlau und Hohwald. Abh Geol Specialkarte Elsass-Lothringen, Strassburg

Salvador H (1987) Stratigraphic classification and nomenclature of igneous and metamorphic rock bodies. International subcommission of stratigraphic nomenclature. Bulletin of the Geological Society of America 90: 440–442

Sawyer EW (1996) Melt segregation and magma flow in migmatites: implications for the generation of granite magmas. Transactions of the Royal Society of Edinburgh. Earth Sciences 87: 85–94

Scott RH (1862) On the granitic rocks of Donegal and the minerals therewith associated. Quarterly Journal of the Geological Society of London 15: 84–86

Sederholm JJ (1907) Om granit och gneiss. Bulletin Commision of Geology Finland 23, pp 110

Sewell RJ, Darbyshire DPF, Langford RL & Strange PJ (1992) Geochemistry and Rb/Sr geochronology of mesozoic granites from Hong Kong. Transactions of the Royal Society of Edinburgh: Earth Sciences 83: 269–280

Shand SJ (1927 &1947) Eruptive Rocks. T. Murby & Co, London, pp 444

Silver LT & Early TO (1977) Rubidium–Strontium fractionation domains in the Peninsular Ranges Batholith and their implications for magmatic arc evolution. (Abstract) Transactions of the American Geophysical Union 58: 532

Singh DS, Chu LH, Loganathan P, Teoh LH, Cobbing EJ & Mallick DIJ (1984) The Stong Complex: a reassessment. Bulletin of the Geological Society of Malaysia 17: 61–77

Skerlie KP & Johnson AD (1993) Vapor–absent melting at 10Kb of a biotite and amphibole bearing tonalite gneiss; implications for the generation of A-type granites, Comment and reply. Geology 21: 89–90

Stephenson PJ (1990) Layering in felsic granites in the Main East Pluton Hinchinbrooke Island, North Queensland, Australia. Geological Journal 25: 227–238

Stone M (1974) A study of the Praa Sands Elvan and its bearing on the origin of elvans. Proceedings of the Ussher Society 3: 37–42

Streckeisen AL (1976) To each plutonic rock its proper name. Earth Science Reviews 12: 1–33

Streckeisen AL & Le Maitre RW (1979) A chemical approximation to the modal QAPF classification of the igneous rocks. Neues Jahrbuch Minerall 136: 169–206

Talbot CJ (1971) Thermal convection below the solidus in a mantled gneiss dome, Fungwi Reserve, Rhodesia. Journal of the Geological Society of London 127: 377–410

Taylor WP (1975) Intrusion and differentiation of granite magma at a high level in the crust: the Puscao Pluton, Lima Province, Peru. Journal of Petrology, Vol 17: 194–218

Thomas HH & Campbell–Smith W (1931) Xenoliths of igneous origin in the Tregastel–Ploumanac'h granite, Cote du Nord, France. Quarterly Journal of the Geological Society of London 88: 274–296

Varlamoff N (1972) Central and West African rare metal granitic pegmatites, related aplites quartz and mineral deposits. Mineralia Deposita 7: 202–216

Vernon RH (1984) Microgranitoid enclaves in granites–globules of hybrid magma quenched in a plutonic environment. Nature 309: 438–439

Vernon RH (1984) Possible role of superheated magma in the formation of orbicular granitoids. Geology 13: 843–845

Vernon RH (1986) K–feldspar megacrysts in granites–phenocrysts not porphyroblasts. Earth Science Reviews 23: 1–63

Vigneresse JL (1990) Use and misuse of geophysical data to determine the shape at depth of granitic intrusions. Geological Journal 25: 227–238

Villaseca C, Barbero L & Herreros V (1998) A re-examination of the typology of peraluminous granite types in intracontinental orogenic belts. Transactions of the Royal Society of Edinburgh: Earth Sciences 89: 113–119

Weaver SD, Adams CJ, Pankhurst RJ & Gibson IL (1992) Granites of Edward & Peninsular, Marie Byrd Land: anorogenic magmatism related to Antarctica–New Zealand rifting. Transactions of the Royal Society of Edinburgh: Earth Sciences 83: 281–290

Whalen B, Currie KI & Chappell BW (1987) A–type granites: geochemical characteristics, discrimination and petrology. Contributions to Mineralogy and Petrology 95: 407–419

White AJR & Chappell BW (1977) Ultrametamorphism and granite genesis. Tectonophysics 43: 4–22

White AJR & Chappell BW (1983) Granitoid types and their distribution in the Lachlan Fold Belt, Southeastern Australia. In Roddick JA (ed) Memoir of the Geological Society of America 9: 21–24

White AJR, Clemens JD, Holloway JR, Silver LT, Chappell BW & Wall VJ (1986) S–type granites and their probable absence in Southwestern North America. Geology 14: 115–118

Whittington HB (1981)The significance of the fauna of the Burgess Shale, Middle Cambrian, British Columbia. Proceedings of the Geologists Association 91: 127–148

Whittington HB & Conway Morris S (1985) Extraordinary fossil biota: their ecological and evolutionary significance. Philosophical Transactions of the Royal Society: London B311: 1–92

Williams IS, Compston W & Chappell BW (1983) Zircon and Monazite U–Pb systems and the histories of the I–type magmas, Berridale Batholith, Australia. Journal of Petrology 24: 76–97

Wilson PA (1975) Potassium/Argon studies in Peru with special reference to the emplacement of the Coastal Batholith. PhD Thesis, Liverpool University

Wilson HE (1972) Regional geology of Northern Ireland. HMSO, pp 115

Winkler HG (1967) Petrogenesis of metamorphic rocks. Springer Verlag, Berlin, pp 348

Xu Keqin, Sun Nai, Wang Desi, Hu Shouxi, Liu Yig Chun & Ji Shouyuan (1982) On the origin and metallogeny of the granites of South China. In Xu Keqin & Tu Guangchi (eds) Geology of the granites and their metallogenic relations. Proceedings of the International Symposium, Nanjing University, Science Press Beijing: 1–3

Zhu Jinchu, Xu Shijing, Shou Huiqun, Shen Weizhou & Liu Changshi (1988) Tin/Tungsten bearing granites in South China. 5th International Symposium on Tin/tungsten granites in Southeast Asia and the Western Pacific. IGCPproject 220 Shimane University, Japan: 262–265

Index